Wood-Based Energy in the Northern Forests

Michael Jacobson • Daniel Ciolkosz
Editors

Wood-Based Energy in the Northern Forests

Springer

Editors
Michael Jacobson
Department of Ecosystem Science and Management
Pennsylvania State University
University Park, PA, USA

Daniel Ciolkosz
Department of Agricultural and Biological Eng.
Pennsylvania State University
University Park, PA, USA

ISBN 978-1-4614-9477-5 ISBN 978-1-4614-9478-2 (eBook)
DOI 10.1007/978-1-4614-9478-2
Springer New York Heidelberg Dordrecht London

Library of Congress Control Number: 2013955039

Printed on acid-free paper

Springer is part of Springer Science+Business Media (www.springer.com)

Contents

Contributors

Jim Babcock Advanced Recycling Equipment Inc., St Mary's, PA, USA

Charles D. Canham, Ph.D. Cary Institute of Ecosystem Studies, Millbrook, NY, USA

John E. Carlson, Ph.D. The Department of Ecosystem Science and Management, The Schatz Center for Tree Genetics, Pennsylvania State University, University Park, PA, USA

Daniel Ciolkosz, Ph.D. Department of Agricultural and Biological Engineering, The Pennsylvania State University, University Park, PA, USA

Alexander M. Evans, Ph.D. The Forest Guild, Santa Fe, NM, USA

Damon Hartley, M.S. Division of Forestry and Natural Resources, West Virginia University, Morgantown, WV, USA

Joshua R. Herr, Ph.D. The Department of Ecosystem Science and Management, The Schatz Center for Tree Genetics, Pennsylvania State University, University Park, PA, USA

Polly Hoppin, Sc.D. School of Health and Environment/Lowell Center for Sustainable Production, University of Massachusetts Lowell, Lowell, MA, USA

Molly Jacobs, M.P.H. Lowell Center for Sustainable Production, University of Massachusetts Lowell, Lowell, MA, USA

Michael Jacobson, Ph.D. Department of Ecosystem Science and Management, Pennsylvania State University, University Park, PA, USA

Barbara B. Kinne Penn State Institutes of Energy and the Environment, University Park, PA, USA

Matthew Langholtz, Ph.D. Environmental Sciences Division, Oak Ridge National Laboratory, Oak Ridge, TN, USA

Jessica Leahy, Ph.D. School of Forest Resources, University of Maine, Orono, ME, USA

Laura Lindenfeld, Ph.D. Department of Communication and Journalism and the Margaret Chase Smith Policy Center, University of Maine, Orono, ME, USA

Bruce Lisle Master of Pellets from the Energex University, Mifflintown, PA, USA

Weiguo Liu, M.S. Division of Forestry and Natural Resources, West Virginia University, Morgantown, WV, USA

Robert T. Perschel New England Forestry Foundation, Littleton, MA, USA

Corie Podschelne Harman Home Heating, Halifax, PA, USA

Adam Stern The Gemstone Group Inc., Wayne, PA, USA

G. Peter van Walsum, Ph.D. Chemical and Biological Engineering, University of Maine, Orono, ME, USA

Jingxin Wang, Ph.D. Division of Forestry and Natural Resources, West Virginia University, Morgantown, WV, USA

M. Clayton Wheeler, Ph.D. Chemical and Biological Engineering, University of Maine, Orono, ME, USA

Introduction and Overview

This volume of papers grew from a Penn State short course on developing wood bioenergy projects, held in November 2011. It was intended as a hands-on course to better identify winning scenarios, avoid costly mistakes, and develop biomass projects that are truly sustainable. The course focused around a bioenergy case study where participants actually developed a biomass project. Given the focused nature of the short course, additional papers for this volume were solicited to provide the reader with a comprehensive snapshot of wood energy in the Northeast. What emerged is a fascinating collection of analysis and discussion on some key issues related to wood energy in the Northern Forest, with special focus on the Northeast United States.[1] The topic is very broad, involving science, engineering, business, and society. The chapters of this volume are similarly diverse, ranging from careful scientific analysis to practical guidance to regional history and so on.

The book's goal is to provide the reader with a sense of the state of woody biomass development in the Northeast region, as well as an overview of the issues in scaling up its production and utilization. Most chapters provide practical hands-on advice for the practitioner, so a key audience is anyone developing a woody bioenergy project.

The book starts with an overview of the state of bioenergy in the Northeast including its availability, economics and environmental aspects of the resource. Then the book shifts to production, harvesting, processing, and uses of bioenergy. Following chapters cover harvesting systems, pellets, residential (heat) and nonresidential—commercial/industrial (heat and power) and liquid fuels. Finally, the book addresses community (e.g., health effects), social, and financial considerations.

Interest in woody biomass is growing, in part because it is a renewable energy source, but also because the technologies and costs for its use have improved dramatically in recent years. From an environmental perspective, wood biomass has several advantages compared to fossil fuel. These include lower carbon emissions,

[1] For the purposes of this book, we will consider the Northeast United States to include the 12 states of Maine, New Hampshire, Vermont, Massachusetts, Rhode Island, Connecticut, New York, New Jersey, Pennsylvania, Maryland, Delaware, and West Virginia.

reduced threats of acid rain, and low particulate emissions if standard emission control devices are used. In addition, using woody biomass can be a tool to help regenerate the forest and improve forest health, biodiversity, and wildlife habitat. From a socioeconomic perspective, locally produced wood biomass strengthens local economies and provides better energy security. Another benefit is that there are multiple and diverse sources of biomass available which can keep costs stable over time. Compared to other renewable energy sources as such as solar and wind, biomass is the only one that can be effectively stored and delivered when needed.

One of the benefits of bioenergy compared to other renewables such as solar or wind energy is its usability in all energy sectors. Chapters 5–9 discuss these uses. Its most common and oldest use in the Northeast is for *thermal* applications, namely wood stoves, pellets, heat for schools, hospitals, and other such community scale institutions. Some facilities are also using biomass for *combined heat and power* (CHP) systems. In addition, the *electricity* sector has existing and proposed wood-fired or co-fired power plants across the region. Finally, and not yet in full commercial production are *liquid fuels*, such as cellulosic ethanol. The potential for expanding the use of wood in all these energy sectors is significant in the Northeast, especially thermal due to the region's high dependence on costly heating oil in the winter months.

In the Northeast, biomass for thermal energy needs (space and process heat) is the most practical use at this point in time and has significant near-term growth potential. The forest products industry has long used wood for heating their facilities, but now its use has grown for residential, commercial, and industrial applications as well. Chapters 6 and 7 focus on residential and nonresidential combustion systems. New heating systems are emerging as are business models to promote their application. But as with all new markets there are limitations such as systems fuel requirements, delivery, and storage issues.

Common sources of woody biomass include whole tree harvest (chips or roundwood), timber harvest residue (e.g. bole chips), mill residue (chips and sawdust), used "ground" pallets, pellets, construction and demolition debris, and dedicated short rotation woody crops (e.g., hybrid poplars and willows) usually cultivated using agricultural practices. Much of the immediate opportunity for wood fuel in the region comes from the timber harvest residues—these residues can include small diameter logs, branches and limbs, bark, needles, and stumps. Typically, logging residues can make up about 25–45 % of the tree's biomass when trees are harvested for sawtimber or pulpwood. Much of the harvesting that occurs in the Northeast, especially unsustainable "high grades" leaves large amounts of logging residues as well as small-diameter or low-value trees.

In some cases loggers are harvesting whole trees and chipping them. Most commonly, chips are usually made from the tops, branches, and nonmerchantable parts of the tree. These are by-products after the logger sets aside the higher valued sawlogs and veneer. Often this biomass is chipped at the landing and sent directly to the processing facility. Some of the logging residues are used in pulp markets, but in areas where the demand for pulpwood has decreased, bioenergy can provide a valuable market for this material. Chapter 2 discusses the availability and potential for biomass from integrated forest operations in the region.

Mill residues are the wood that remains after boards and other wood products are manufactured. This includes sawdust as well as chips or shavings ("slab wood" is not as common anymore). This wood is usually dryer than whole tree chips, and sawdust may be useful only in certain fuel handling systems. Pellets from sawdust have become one of the fastest growing biomass markets, both for domestic use and for the relatively new and expanding export market worldwide. Chapter 8 discusses the history and development of the pellet market in the region. Although they are more expensive than wood chips or roundwood, important benefits of pellets are the ease of handling and the consistency of size and moisture content. These systems are discussed more fully in Chaps. 6 and 7. Construction and municipal waste, another important source of woody biomass, is plentiful in the region, especially in urbanizing areas where disposal costs make it an inexpensive fuel source but also a concern in terms of toxics and impurities.

One of the main drivers of wood energy in the region is the lower fuel price as compared to fossil fuels, particularly in thermal systems. Although the market for woody biomass is still young, it parallels that of the pulp wood markets. Some areas have competitive markets, especially where pulp markets are active. Prices are highly localized and depend primarily on harvesting and transportation costs. Factors influencing price include geography, delivery method, and transportation network. Chapter 2 discusses the economic models of feedstock availability across the region. Another economic consideration is the fragmented supply chain from the landowner to the end user that includes the forester, the loggers, the truckers, and the processors. Developing an efficient, effective supply chain is an important challenge facing the industry at present.

On the societal side, wood energy in the region requires dealing with landowners who often think mainly about nontimber benefits, and need convincing that biomass harvests can be a sustainable component of responsible forest stewardship. On the other hand, when profit is a major consideration, one finds that prices for biomass (on the stump) are often very low. To make the biomass profitable its harvest must usually be coordinated with other highly valued timber removals to sway a landowner to sell biomass. One interesting factor that has arisen recently in the region is the newfound abundant supply of shale-deposit natural gas. It remains to be seen how this local energy boom will affect landowner perspectives and interest in renewable energy. Chapter 11 discusses some of these social considerations.

There are also health and environmental concerns related to the use of wood for energy. The top health issue is undoubtedly concern about air emissions, especially particulate matter, from burning biomass for heat, which is discussed in Chap. 10. The key environmental issue revolves around proper forest management to ensure sustained long-term site productivity. Harvesting biomass for energy could impact soil nutrients, organic matter and soil moisture-holding capacity. Nitrogen and other essential plant elements are abundant in twigs and foliage so that harvesting all above-ground biomass could theoretically remove a large proportion of nutrients. Several states in the northeast have developed Forestry Best Management Practices (BMPs) for biomass harvesting guidelines to address these concerns. This is discussed more in Chap. 1. Lastly, the complex question of carbon emissions and wood energy is an area of interest for policy developers that must be addressed if our picture of wood energy is to be complete (see Chap. 3). Interestingly enough, health and environmental concerns tend to be largely dependent on how wood energy is implemented, rather than if it is implemented—methods and technologies are expected to have a great effect on how wood energy impacts the regional ecosystem.

Finally, a biomass project requires financing and dealing with risk. Perhaps the key risk is ensuring constant and stable supply of feedstock. Although it is well recognized that most of our region has plentiful supplies of wood, availability, cost, and intangibles, such as transportation costs, are an issue. Making sure one has enough suppliers, solid contracts with suppliers, enough storage capacity in the event of shortfalls, adds costs but also mitigates risk. Economic questions are dealt with throughout the book, especially in Chap. 2, whereas Chap. 12 discusses financing.

The following is a short synopsis of each chapter.

Chapter 1 sets the stage and context for developing wood-based bioenergy projects in the Northeast. The authors describe the forest condition, extent of wood in the region, and lay out some issues and constraints to wood energy development. The region is relatively large and forests are diverse. Almost three quarters of the forest are in 4 of the 12 states in this region. Given the diverse forest types one finds a varied forest products industry. Generally the region's industry specializes in high-quality saw and veneer wood products. Maine is an exception with its very strong pulp market. Understanding the current wood product market is critical to bioenergy development since a key question is how wood for bioenergy complements or competes with other forest products.

Local considerations in project development are critical. Since wood energy is sourced locally it takes pressure off the need to rely on imported supplies. However, this creates its own issues with respect to local regulations and environmental impacts. Chapter 1 explains how state and even local policies have evolved to address harvesting guidelines for wood energy.

The region has abundant supplies of low-use wood and the forests have a positive tree volume growth-to-drain (removal) ratio. However, as Chap. 1 discusses, perhaps the biggest question in the Northeast is how much biomass is there that is practically available? The region is heavily populated with 21 % of the country's population and only 6 % of the land area. On the one hand the high population and urban areas make for high energy demands (currently mainly in the form of coal and nuclear) but from which biomass could contribute. On the other hand, population pressures lead to urbanization/deforestation and forest fragmentation. The forests in our region are already highly fragmented with millions of private landowners owning small forested parcels. Developing bioenergy projects will require working with these landowners who in many cases prefer esthetic, wildlife and other amenities to harvesting timber.

Chapter 2 addresses the key economic questions of feedstock supply availability and at what price. The chapter shows results from agricultural simulation models that project supplies by 2030 given a variety of farmgate prices for the Northeast region. Key to any question of availability are production costs and competition from alternative markets. The manner in which supply changes with price allows policy makers and producers determine feasibility of alternative opportunities. The models are based on the Department of Energy's Billion Ton Report Update of 2011. The two main sources of woody feedstock in the region are projected to be forest resources - such as, thinnings (thinnings and logging residues), and short rotation woody crops (SRWC). The SRWC comes from nonforested land and the paper explains assumptions for which land is available for SRWC versus other agricultural crops and pasture. Baseline projections suggest an increase in biomass production from a current 4.3 dry million tons at $22 per ton over 25 million tons at $88 per ton in 2030. SRWC will become a significant component of the feedstock by 2020. The model estimates in 2020 at a price of $66 per ton, there will be about 13 million tons of woody biomass available in the Northeast of which 7.5 comes from SRWC. Most of the material will initially come from forest rich states such as Maine, Pennsylvania, and West Virginia while SRWC will come from agriculture rich states such as New York and Pennsylvania.

Chapter 3 builds on the ecological issues raised in Chap. 1 but focuses on carbon. Harvesting woody biomass will have significant impacts on forest carbon stocks and flows. It is clear that our northern forests as a whole are a carbon sink (growth exceeds removals and mortality). But there are important questions as we ramp up production of bioenergy. What will happen to forest carbon levels in the event of increased biomass harvesting? How much can we increase harvest while maintaining stable sequestration rates? Using stochastic models this chapter addresses these questions. It first describes forest carbon dynamics, mainly due to harvesting. Even setting a "baseline" level of carbon storage is very controversial. Although northern forests are currently net carbon sinks it is not a given it will remain that way, even without biomass harvests. However, carrying out "business as usual" harvest regimes could see a concomitant increase in sequestration since we mostly carry out partial cuts. However increasing harvest levels could have negative implications for carbon levels in forests. Modeling, of course is fraught with assumptions and the author points out that forest dynamics are not "inherently steady state." The chapter also discusses implications for net carbon impacts taking into account offset from reduced fossil fuel use.

Chapter 4 discusses the state of tree improvement and breeding both for forest trees and short rotation woody crops (focusing on willow and hybrid poplar). For viable bioenergy production across the supply chain, perhaps the most important factor is consistently high yields of biomass. The key goal of breeding is to increase yields, especially on marginal lands, but other traits are also examined depending on the end use, such as resistance to pests and diseases. Depending on the end use, traits can be bred for high energy/lignin content for thermal applications or high cellulose/sugar content for liquid biofuels. Due to the long rotation ages of forest trees the focus of current research is on SRWC, where new improved cultivars can be bred in terms of years instead of decades. The chapter discusses all the relevant and latest tools and techniques from traditional breeding to genomics and genetic engineering.

Chapter 5 addresses the issue of optimizing biomass harvest systems. Harvest systems include the processes used to fell and collect trees and transport them to the processing facility. The success of a bioenergy project can hinge on selecting harvest systems that are both financially and environmentally effective. The chapter discusses productivity and costs of various harvesting systems and provides a detailed overview of the variety of harvesting systems from "wood energy bundlers" to "coppice headers." This chapter is especially useful to practitioners and researchers in showing how to measure harvest productivity and its associated costs. This ranges from basic "time and motion" measurements to integrated Global Positioning Systems (GPS) approaches. The types, uses, and pros and cons of using economic models to analyze component costs of the harvesting systems are also discussed. One key lesson is that increased harvest efficiency must be weighed against potential site impacts such as future productivity and environmental degradation. Finally, the chapter lays out a variety of scenarios showing how delivered costs vary for different harvesting and transportation systems.

Chapter 6 focuses on wood used in residential heat. Many take this sector of woody heat for granted, but it plays an essential role in future development of biomass markets. The most common use of wood biomass in Northeast is for heating homes. Over one-half million homes in the Northeast use wood heat and there seems to be a renewed interest as the costs of alternative fuels increase and improved and more efficient stoves emerge on the market. For many homes it comes down to tradeoff between cost savings with wood versus oil (or other fossil fuels) and the convenience factor, i.e., not having to deal with storage and handling issues. Wood fuel's cost is the least expensive on a gigajoule (GJ) basis and least volatile in terms of price fluctuations. The authors discuss the history and use of wood heat in the Northeast, noting its decline since the advent of fossil fuels in the late 1800s. The chapter details the types and technologies of different burning devices. In recent years, the industry has responded to development of new technologies and tightening of EPA regulations, making stoves much cleaner and more efficient. Different types of wood stoves, pellet stoves and outdoor boilers are options for homes. For example, pellet stoves are more efficient and consume less wood compared to other devices. The authors end their chapter by discussing prospects for wood heat in the future, which basically comes down to tradeoffs in cost of other fuels, health issues, and convenience.

Chapter 7 looks at the commercial side of heat and electricity from woody biomass. Although wood heat for homes is most common, there is growing interest in replacing institutional oil boilers with wood boilers to save on fuel costs. As a result schools, office buildings, hospitals, and manufacturing plants are examples of places where wood heat and even Combined Heat and Power (CHP) systems can work economically. The chapter describes the different type of system configurations from fuel storage and handling to combustion design and emission controls. System design varies quite widely and it is important especially with respect to type of and quality of woody fuel used. If electricity production is a consideration, the type of generator is an important consideration. Key issues are automation of the system, type of fuel, and installation costs. Retrofitting existing facilities can be costly. Determining payback and financial returns of a project is important as these costs, as the authors say can "creep" up. Perhaps the most important part of designing a project is ensuring good system performance from whatever design is selected. Performance implies factors such as thermal and electrical efficiency (how much fuel is actually available for heat for electricity) and emissions performance. Finally the chapter discusses emerging technologies such as gasification and Organic Rankine Cycle systems.

Chapter 8 presents a historical perspective of the pellet market in the Northeast. Written by one of the early developers of the market, its reads like a biography of sorts. Scattered throughout are great anecdotes about the nascent and early days of wood pellet production. Traditional wood energy has been around a long time but it is pellets are more recent and beginning to influence the market. The early days saw wood stove proliferation after the 1973 oil embargo, only to be followed by EPA standards dramatically reducing the number of wood stove manufacturers. However pellet stoves, because of their high air-to-fuel ratio were exempt and so the rest is history. What's interesting is that the first markets on the east coast were mainly for institutional and industrial customers, whereas today the residential market dominates. Over the years there were many ups and downs in the pellet market with competition coming from gas stoves, low oil prices, and pellet storage shortages where demand exceed supply. New capacity was installed after hurricane Katrina, but in the last few years millions of tons of new capacity are being built for export to Europe. However, the US market is also rapidly developing as the pellet fuel industry is becoming certified and commercial, and institutional-type consumers are realizing the fuel savings from central heat, using pellets delivered in bulk, which the author suggests will soon outpace and dwarf the current bag market.

Chapter 9 focuses on methods for converting wood to liquid fuels. This of course is of particular interest given the potential carbon emission reductions and associated benefits of replacing fossil fuels. The chapter considers all the available platforms for converting wood to liquids. The state of liquid biofuels is at the cusp of commercial production and it is timely to look at what methods are suited to different woody feedstocks. The technologies and science are quite complex but the authors lay out a concise, orderly description of the methods. The authors discuss mainly sugar (hydrolysis and fermentation) and pyrolysis platforms but mention briefly gasification and other platforms such as carboxylate platforms. For each platform the authors provide detailed information about the method of feedstock conversion, pretreatment (if needed), the type product (i.e., ethanol, and drop in fuels) and which forest resources are most applicable or preferred to that particular method/platform. Key issues also include conversion costs, access and availability of resources, and technological limitations. One important caveat is that the scale of the processing facility is limited by resource availability, and even the largest corn ethanol facility is an order of magnitude smaller than an average oil refinery (400 million liters per year vs. 6.8 billion liters per year). It is interesting to note that a

400 million liter (~100 million gallon) per year biorefinery corresponds to the largest commercial pulp mills, with a capacity of 3,000 dry tonnes/day of wood delivered.

Chapter 10 provides a succinct summary of the knowledge about wood combustion emissions and impacts on health. The authors focus on nonresidential scale facilities such as schools and hospitals (see Chap. 7). The chapter first looks at the evidence of exposure to emissions. The EPA lists over 100 chemical pollutants in wood combustion emissions. Of major concern is particulate matter, especially particles less than 2.5 μm in size. Other important pollutants are carbon monoxide, and polycyclic aromatic hydrocarbons. Key issues for examining pollutants and their health impact are the type of fuel used, the moisture content, the combustion technology, how the system is operated, and the emission controls used. Each of these five issues is clearly discussed and elaborated in the chapter. The evidence linking health concerns and the types of populations more at risk is discussed. Specific information needed to make decisions about health impacts from wood burning emissions is noted which leads to a discussion on policy issues and recommendations for protecting public health. The authors focus on air-quality regulations and roles of states and EPA in setting requirements. Recommendations include providing incentives for cleaner burning systems, more research into health impacts, and consistent and standardized regulations across the region.

Chapter 11 provides insight into an important cog in the biomass wheel, the social issues. Most of the book deals with economic- and engineering-type issues, but for biomass to grow, understanding societal concerns from using biomass for energy must be addressed. The chapter focuses on ways we can take the science of bioenergy and link it to practice to meet social needs. The authors call it "knowledge to action" where the researcher's knowledge and "supply" is matched with the stakeholders "demand." Stakeholders include landowners, communities, industry, and government. The paper assesses stakeholder's needs and knowledge about the emerging bioenergy sector in the Northeast. They examine barriers, opportunities, and research needs by interviewing a diverse set of people involved in the bioenergy sector. Key barriers include economic issues and public perception of the industry. Key opportunities and needs are developing sociocultural capacity, especially with growers and landowners and developing viable feedstock production and processing systems that are sustainable. Significant points emerging from this survey of stakeholders is the importance of interdisciplinary approaches, the need for early identification of issues, ability to be adaptive as the project develops, and for trustful collaboration. As one the interviewees stated, the biomass and technology is available, the key is "showing people that it can be done sustainably and moving forward."

Chapter 12, the final chapter, although the author suggests that it is the first step not the final step, is all about financing biomass projects. That is true—without a clear, solid, and doable financial plan and its implementation—the project will go nowhere fast. It may sound simple i.e., raising capital and using cash flow to pay off debt, but it is far from that. Focusing on project finance, as opposed to corporate finance, the paper delves into really critical issues and decisions any biomass project needs to consider. What is important is the financing process which includes the project structure, its documents, and how to attract financing. Project agreements, of which there are many, include consents and contracts, which are the legal foundation of the project. Then there is due diligence to ensure that the investor will finance a project, including a package that supports and justifies the project. The variety of loan types is another area which requires some thought since many biomass projects are unproven hence needing unique approaches to garner finance. The chapter clearly explains all the factors to consider and ends with some unique things relative to a biomass project, which include ensuring an ample supply of feedstock, addressing any environment-related issues with sourcing feedstock, and making sure the project is performing to specifications after construction. Accounting for unexpected and usually rising costs over the life of the biomass project for operation and maintenance is also a crucial consideration in financial planning.

The editors gratefully acknowledge the hard work and dedication of the many authors who contributed to this volume and look forward to the continued work that is being done in this important area. We also like to acknowledge the Northeast Woody/Warm-season Biomass Consortium (NEWBio) project funded by the USDA-NIFA. It is our sincere wish that this book will contribute to the better understanding of, appropriate use, and wise care of the outstanding forests that we are fortunate to have in this region.

University Park, PA

Michael Jacobson
Daniel Ciolkosz

Chapter 1
Forestry in the Northeast

Alexander M. Evans and Robert T. Perschel

1.1 The State of Forests in the Northeast

Forests are the dominant land use in the Northeast United States, and as such are of significant interest as a source of fuel to meet energy needs in a sustainable, renewable manner. Unlike fossil fuels which are often shipped thousands of miles from source to use, the production of forest bioenergy is often a much more local concern. The biogeographic setting determines growth rates and in turn the sustainable supply of a feedstock. Local and regional economics have a greater influence on forest biomass prices than do the global markets which dictate the prices of fossil fuels such as oil. Local climate influences the quantity and type of bioenergy best suited for heating and cooling applications. Even at the policy level, a patchwork of municipal and state policies promotes, retards, or regulates forest bioenergy. This chapter provides an introduction to the ecological, social, and economic context for forest bioenergy in the Northeast.

1.1.1 Geography of the Forests in the Northeast

There are numerous ways to delineate the northeast region of the United States, from ecological zones to administrative boundaries. Here our goal is to describe an area that is similar enough in ecological, social, and economic dimensions to permit generalizations but also large enough to allow for regional relevance. Nearly all

A.M. Evans, Ph.D. (✉)
The Forest Guild, 2019 Galisteo St., Suite N7, Santa Fe, NM 87505, USA
e-mail: zander@forestguild.org

R.T. Perschel
New England Forestry Foundation, 32 Foster Street, Littleton, MA 01460, USA
e-mail: rperschel@newenglandforestry.org

M. Jacobson and D. Ciolkosz (eds.), *Wood-Based Energy in the Northern Forests*,
DOI 10.1007/978-1-4614-9478-2_1, © Springer Science+Business Media New York 2013

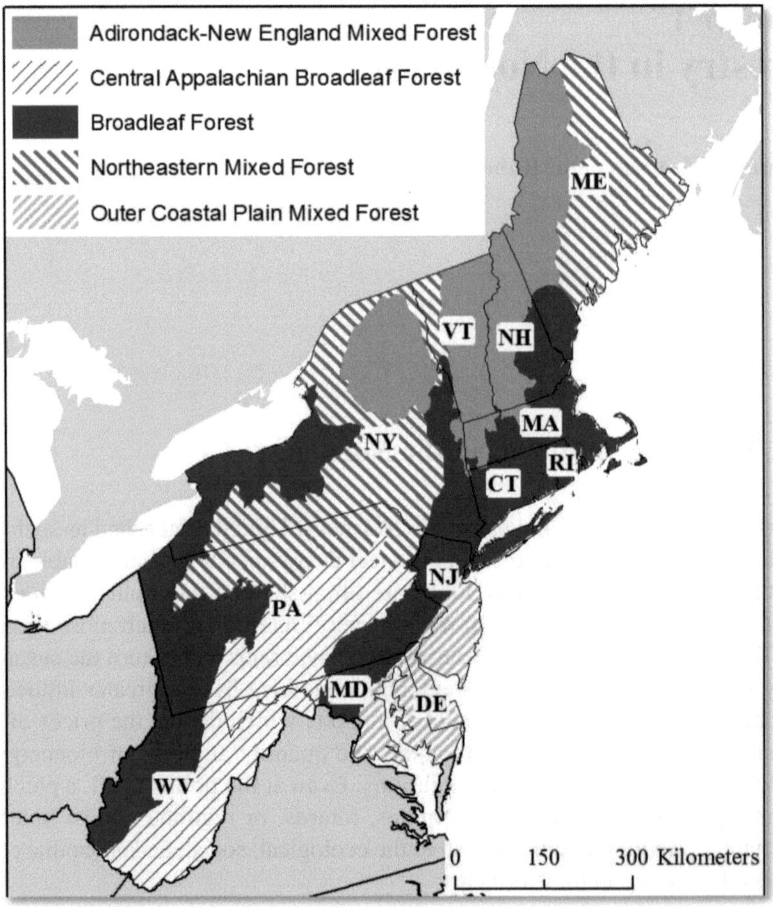

Fig. 1.1 Map of northeast states and forests within the northeast ecological provinces

definitions of the Northeast include the six New England states: Maine, New Hampshire, Vermont, Massachusetts, Rhode Island, and Connecticut. The US Environmental Protection Agency's region 1 includes only the New England states, while the Natural Resource Conservation Service has a more expansive Northeast region. Importantly for discussions of carbon and climate change, the New England states are joined by New York, Pennsylvania, Delaware, Maryland, and New Jersey in the Northeast Regional Greenhouse Gas Initiative (RGGI). Where possible our administrative definition of the Northeast follows the US Forest Service's definition of the Northeast which includes the six New England states, New York, Delaware, Maryland, New Jersey, and West Virginia (Fig. 1.1).

Beyond state boundaries, delineating ecological boundaries is no less difficult. Using the US Department of Agriculture's ecological provinces the Northeast region encompasses four ecological provinces including Northeastern mixed forest,

Adirondack-New England mixed forest-coniferous forest, Eastern broadleaf forest, and Central Appalachian broadleaf forest (McNab et al. 2007). Based on the similarity of forest types this discussion is relevant to forests as far south as Georgia and as far west as Minnesota.

1.1.2 *Climatic and Physiographic Setting*

The Northeast's climatic and physiographic setting drives forest processes and composition. Over the last century, the average annual temperature in the Northeast was 8 °C (46 °F). The winter temperatures average was –4.3 °C (24 °F) while summer temperatures average was 19.6 °C (67 °F) (National Climate Data Center 2012). The prevailing wind direction, from west to east, creates a continental climate except for coastal areas moderated by the Atlantic Ocean (Barrett 1980a). On average over the past century, the region received 104 cm (41 in.) of precipitation which tended to be evenly distributed throughout the year (National Climate Data Center 2012). However, the Northeast is becoming hotter and wetter (Hayhoe et al. 2008). The largest temperature increases have come during the winter, which has warmed at a rate of 0.70 °C (1.3 °F) per decade over the last 35 years (Hayhoe et al. 2007). In the Northeast, the average annual precipitation has increased by 9.5 mm (0.4 in.) over the last century (Easterling 2002, Hayhoe et al. 2007). Very heavy daily precipitation has also increased in the last century (Easterling 2002), and the decrease in the percent of precipitation the Northeast receives as snow has been most notable in northern and coastal areas (Huntington et al. 2004).

The tallest mountain in the Northeast is New Hampshire's Mount Washington at 1,917 m (6,288 ft), but much of the region is set on upland plateaus between 150 and 460 m (500 and 1,500 ft). The Appalachian Mountains are geologically complex and run through the Northeast, including the Longfellow Mountains of Maine, White Mountains of New Hampshire, the Green Mountains of Vermont, the Adirondack and Taconic Mountains of New York, and the Alleghany Plateau region of Pennsylvania and West Virginia. An eastward extension of the Allegheny Plateau forms the Catskill Mountains of New York, while south and east of the plateau, extensive compressional folding of the bedrock is responsible for the complex ridge and valley topography of central Pennsylvania and eastern West Virginia. The region's main rivers, including the Susquehanna, Potomac, Delaware, Hudson, Connecticut, Merrimack, Kennebec, and Penobscot, form wide valleys separating these highlands. The Ohio River and its tributaries cross West Virginia and western Pennsylvania from the headwaters in the Alleghany Mountains.

The Laurentide Ice Sheet covered most of the region until about 12,000 years before the present and left rocky moraines, glacial erratics, kettle ponds, and deep till soils as it retreated. Glaciation created young soils which vary considerably across small spatial scales. The soils of much of the northern portion of the Northeast (Adirondacks, White and Green Mountains, and Maine) are spodosols, acidic and infertile. Inceptisols (young, more fertile soils) dominate much of the southern New

England and southern New York. The Laurentide Ice Sheet never reached southern Pennsylvania, West Virginia, southern New Jersey, Maryland, or Delaware, so in this southern portion of the region there is a higher percentage of older soils. The ultisols found here are old and strongly weathered, but in some locations such as the piedmont region of Maryland and south-central Pennsylvania they are often deep and highly productive.

1.1.3 Forest Types

Major forest types in the region are spruce-fir (*Picea sp.—Abies sp.*), northern hardwood (Eyre 1980), oak-hickory (*Quercus sp.—Carya sp.*) or transitional hardwood forests, and the mixed-mesophytic forests. Spruce-fir forests dominate the inland areas of Maine as well as the mountain tops of the northernmost portions of New York, New Hampshire, and Vermont. Remnant spruce-fir forests can also be found on the mountain tops south into West Virginia. Northern hardwood forests are dominated by maple (*Acer sp.*), beech (*Fagus grandifolia*), and birch (*Betula sp.*) and cover lower elevations and southern portions of Maine, New York, New Hampshire, Vermont, and the northern portion of Pennsylvania. Northern hardwood forests also include conifers, e.g., hemlock (*Tsuga canadensis*) and white pine (*Pinus strobus*), in the mixture (Westveld 1956). The oak-hickory forests are the transitional forest type between the northern hardwood forests type and the mixed mesophytic forest that dominate further south (Westveld 1956). Oak-hickory forests are found on the glaciated inceptisols of southern New England and include a mix of species from the north such as birches, beech, and maples and those from the south such as hickories, ash, and tulip poplar. The inter-mixing of species becomes even more pronounced in the mixed mesophytic forests further to the south. Because these forests acted as a refuge during the last glaciation and because of the great variation in topography, the mixed mesophytic forests are some of the most biologically diverse temperate regions of the world (Loucks et al. 1999). Mixed mesophytic forests contain a variety of magnolias, oaks, hickories, elms, birches, ashes, basswood, maples, black locust, pines, black walnut, tulip-poplar, blackgum, hemlocks, black cherry, and beech.

Of course there are many forest stands that do not fit perfectly within the major forest types described above. For example, pine-dominated forests are found in the coastal areas of Maine and New Hampshire and much of central Massachusetts. These pine forests tend to occupy sites with coarse-textured, well-drained soils (Barrett 1980a). There are also rare forest types that have become the focus of conservation efforts such as white-cedar (*Chamaecyparis thyoides*) swamps. While many of these rare forest types are protected from harvesting, biomass removals for energy may be a crucial restoration tool in fire adapted communities. For instance, pine barrens, forests dominated by fire adapted pitch pine (*Pinus rigida*), may need to have trees removed as the first step in the restoration process. Because humans have reduced the frequency of fire in these forests, more trees have grown up and

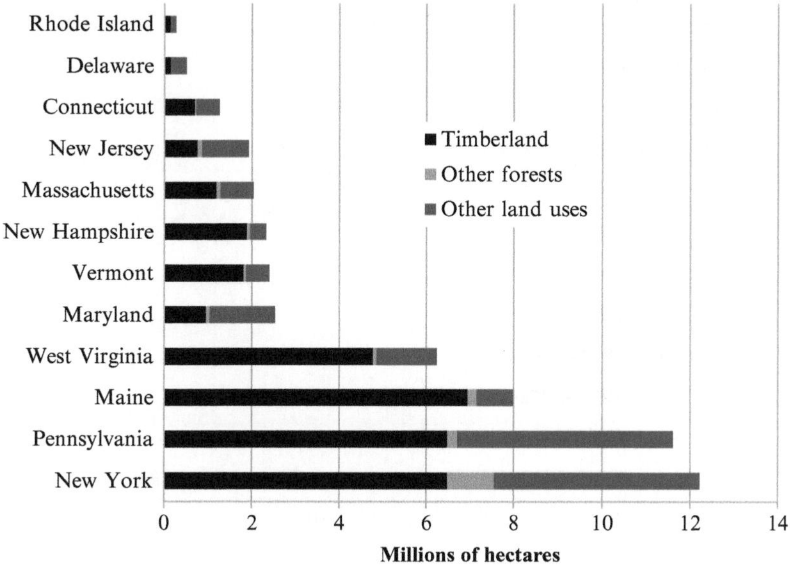

Fig. 1.2 Land cover by state from Smith et al. 2009

they have become unnaturally dense or converted to a closed canopy forest type. Biomass needs to be removed to reduce fuel loads in an effort to protect human life and property, and to perpetuate fire-dependent pine barrens ecological communities (Jordan et al. 2003, Brose and Waldrop 2006). The largest areas of pine barrens are in New Jersey and southeastern New York, but there are discontinuous examples of the forest type on outwash fans and dry ridgetops throughout the region.

1.1.4 Forestry and Forest Products in the Northeast

The Northeast has about 34 million hectares of forests (84 million acres) or about 67 % of the total land area (Smith et al. 2009). Northern states, particularly New Hampshire and Maine, have a greater percentage of land in forest (84 and 89 % respectively) than states with higher population densities such as Delaware which is only 31 % forested (Fig. 1.2). The vast majority, 94 %, of the forest land in the Northeast is timberland, areas where commercial timber could be produced. Reserve forests are a tiny percentage of the landscape except in New York where the Adirondack Park excludes 13 % of the state's forests from harvesting. Similarly, planted forests make up a very small percentage of the timberlands (3 %) in the Northeast in contrast with other regions such as the Southeast where 25 % of timberlands are planted forest (Smith et al. 2009).

About half of the timberland in the Northeast can produce at least 3.5 cubic meter per hectare per year (50 ft^3/ac/year) while the rest of the timberland is less

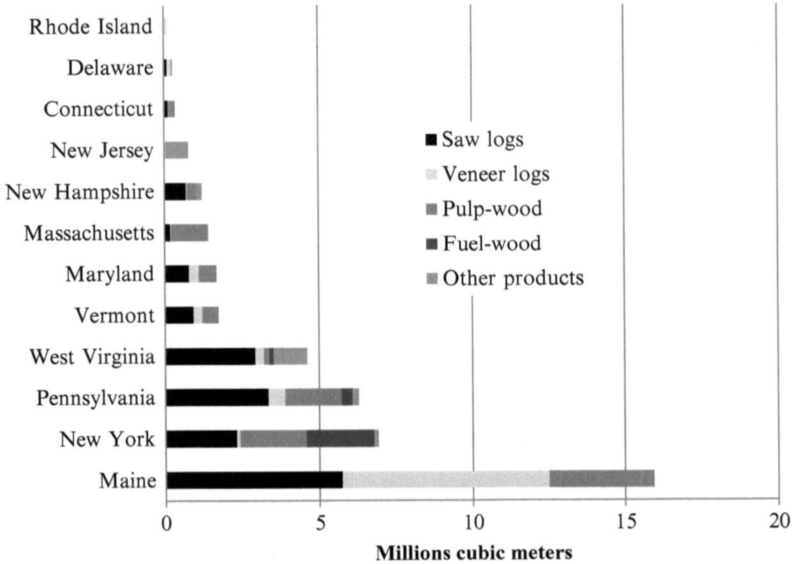

Fig. 1.3 Wood products from the Northeast in 2006 from Shifley et al. 2012

productive (Smith et al. 2009). Maine and West Virginia are the only states with more than a million hectares of timberland than can produce at least 5.9 cubic meter per hectare per year (85 ft³/ac/year). A little over half of the timberland in the Northeast is between 40 and 80 years old and notably, only 6 % is over 100 years old (Smith et al. 2009). However, a recent analysis highlights the difficulty of assigning a single age in forests that are often partially harvested (Canham et al. 2013). Much of the region is comprised of second- or third-growth forest that is yet to reach late seral stages (Irland 1999). Uneven-aged forests make up only 4 % of Northeast timberlands or about 1.4 million hectares (3.5 million acres). The majority of uneven-aged stands are in the Northern hardwoods forests. About 6 % (or 735,000 ha [1,818,00 ac]) of Northern hardwood forests are uneven aged (Smith et al. 2009). Northern hardwoods are well suited to uneven-aged management because many of the commercially important species are shade tolerant.

In the Northeast, net annual growth of growing stock on timber land was 92 million and 33 million cubic meters were harvested in 2006 (Smith et al. 2009). Hardwoods made up 74 % of the growth and 68 % of the harvest in 2006. Of the 33 million cubic meters harvested in 2006, 95 % came from private lands (Smith et al. 2009). In 2006 the Northeast produced about 10 % of the total U.S. roundwood products, but about 25 % of the veneer (Fig. 1.3). Maine has the largest production of sawlogs (5.7 million cubic meters, 203 million cubic feet), veneer (6.8 million cubic meters, 239 million cubic feet), and pulp wood (3.4 million cubic meters, 121 cubic feet). The next largest producer, New York, produced about a third of Maine's volumes (Shifley et al. 2012). About 2.7 million cubic meters of fuel wood, a category particularly relevant to bioenergy production, was produced in the Northeast in 2006 (Shifley et al. 2012). Reports on forest products are beginning to report biomass chip or bioenergy harvests. For example in Maine, biomass chip harvests increased from 0.9 to 2.7 million cubic meters (0.4–1.2 million cords) between 2000 and 2008 (McCaskill et al. 2011). In 2009, the Northeast accounted for about 24 % of the wood pellet production capacity or about 1 million metric tons (Spelter and Toth 2009). In the Lake States the increase in demand from pellet manufacturers is beginning to replace declining demand for lower-quality roundwood from the paper industry (Luppold et al. 2011). Another important product from a bioenergy perspective that can be lost in national harvest assessments is landowner-harvested firewood. In the Northeast about 34 % of family forest owners say they harvest firewood from their property (Roper Public Affairs 2006).

Based on a recent study, the median size of woodshed for softwood and hardwood mills in the Northern New England are about 4,200 and 8,700 km^2 (1,600 and 3,400 mi^2) respectively (Anderson et al. 2011). Areas of high hardwood procurement pressure based on the overlapping woodsheds for existing mills was highest in Herkimer County, New York, central Vermont, and western Maine. In contrast, the same analysis suggests that the areas of greatest softwood procurement pressure were Somerset and Piscataquis counties, north of Moosehead Lake in Maine with significant but light pressure in northern and southwestern New Hampshire (Anderson et al. 2011).

1.1.5 Land Ownership, Conservation, and Fragmentation

Eighty percent of the 34 million hectares of forest land in the Northeast is privately owned (Fig. 1.4). The percentage of private ownership is greatest in Maine (94 %) and is lowest in New Jersey (62 %) and Pennsylvania (71 %). The high percentage of private ownership in Maine is due to a large corporate timber base (over four million hectares [ten million acres]) (Smith et al. 2009). The ownership of forest land in the Northeast has changed significantly in recent decades. One trend has been turnover of land forest holding in the Northern Forest, where 23 million acres were involved in land sales between 1980 and 2005 (Daigle et al. 2012). The change in ownership has generally involved a divestment of timber or wood products companies and an increase in the forest holding of Timber Investment Management Organizations (TIMOs), Real Estate Investment Trusts (REITs), and land trusts. The change in ownership from wood products companies to TIMOs and REITs means that management and silvicultural decisions are not as directly tied to mills' needs. The ownership change also increases the potential for greater land ownership turnover and forest fragmentation. However a recent report from North Carolina highlights how TIMOs and REITs can act as partners in protecting forest land and preserving working landscapes (Weinberg 2012).

On average, forest land owners in the Northeast own about 10 ha (26 acres), but in Southern New England and the mid-Atlantic states the average forest ownership is

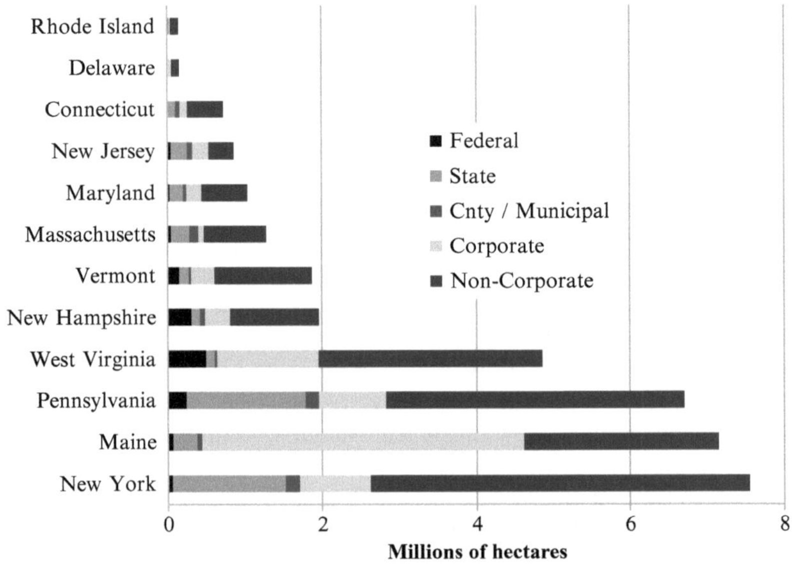

Fig. 1.4 Forest land by ownership from Smith et al. 2009

less than 4 ha (10 acres) (Smith et al. 2009). There are about 2.7 million nonindustrial private forest (NIPF) owners. They own 21 ha (23 acres) on average and that makes up about 73 % of the forest land in the Northeast and (USFS 2012). West Virginia, New Hampshire, and Pennsylvania have the largest federal ownership in the Northeast (434,000; 291,000; and 201,000 ha respectively) because of the National Forests in those states. Pennsylvania and New York have much more state forest land than any other state in the region at about 1.5 million hectares each.

Much of southern New England was cleared for agriculture in the early nineteenth century, leaving less than 1 % of the forest cover in an old-growth condition (Cogbill et al. 2002). Since then, the long-term trend in the Northeast during the twentieth century was one of increasing forest cover (Smith et al. 2009). Forest cover grew from a low of about 44 million acres at the turn of the twentieth century to 70 million acres 100 years later (Fig. 1.5).

This long reforestation process has started to turn around as forest cover (total forested hectares) has recently started to decline again (Drummond and Loveland 2010). In the Northeast Highlands ecological zone forest declined from about 85 % forested to 81 % forested from 1973 to 2000. During the same period the Northeastern Coastal Zone that covers most of southern New England declined from 51 to 47 % forested (Drummond and Loveland 2010). Predictions from the North East State Foresters Association suggested that as much as 368,000 ha (909,000 ac), or about 2 % of forest land, could convert from forest to other land uses in Maine, New Hampshire, Vermont, and New York by 2050 (Sendak et al. 2003). These predictions

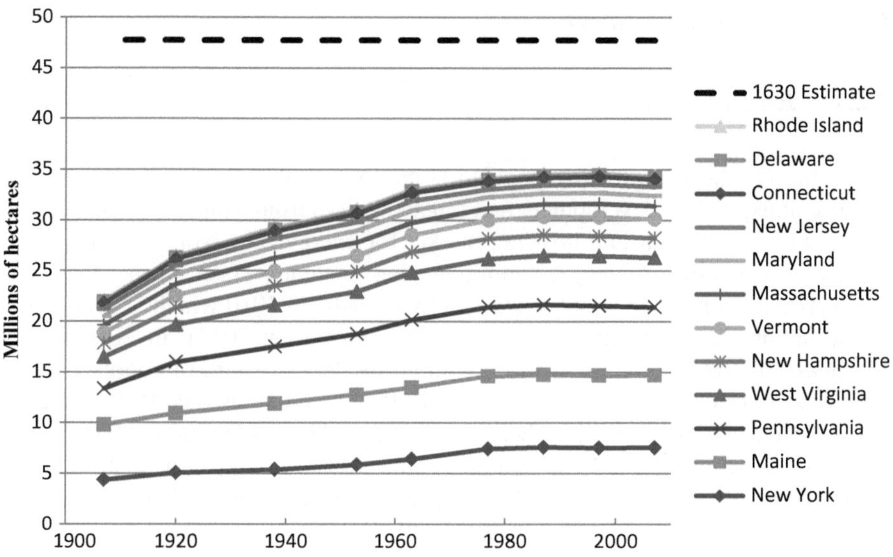

Fig. 1.5 Forest cover over time in the Northeast from Smith et al. 2009

for the Northeast are similar to national scale models for land use change (Theobald 2010, Radeloff et al. 2011).

A key driver of fragmentation of forested landscapes in the Northeast is the high density of human population in the region. More than 84 % of the Northeast region's 55 million people live in urban areas mostly along the coast (US Census Bureau 2010). Though the Northeast has not grown as fast as some regions of the country, between 2000 and 2010 its population increased by 1.7 million people. The increase in population no doubt helped drive the development that has occurred in formerly rural or forested areas causing landscape fragmentation (Egan et al. 2007). Other drivers of development in the forests of the Northeast include deconcentration of housing and amenity-driven population growth in rural areas (Hammer et al. 2009). In the Catskill-Delaware watersheds owners identified property taxes, aging, lack of interest in forest ownership among family members, and personal circumstances as concerns in owning forestland and reasons for parcelization (Stone and Tyrrell 2012).

Development in rural communities and forested area means more than just loss of the trees cut down to make room for houses or commerce. Forest fragmentation threatens forest processes and wildlife habitat (Stein et al. 2010). Declines in forest area and the related forest fragmentation are magnified by the expansion of exotic, invasive plants and insects, which are encouraged by housing and other development (Londré and Schnitzer 2006, Gavier-Pizarro et al. 2010). There also tends to be a reduction in the amount of harvesting activity as nearby land is converted to nonforest uses (McDonald et al. 2006).

1.2 Economic Context

1.2.1 Energy Use

While the Northeast's growing population can pose a threat to forest ecosystems because of expanding development and fragmentation, from a bioenergy perspective the growing population is a growing market for energy. The Northeast (using the Energy Information Agency's definition that includes states from Maine to Pennsylvania) consumed about 14.45 EJ (13.7 quads) in 2010 which is about 14 % of the national total (Energy Information Administration 2012). In 2010 for the Northeast, the transportation section required the largest share of energy (30 %), followed by residential (27 %), commercial (25 %), and industrial (18 %). In 2010 38 % of the electricity in the Northeast was generated by nuclear power followed by coal (28 %), natural gas (24 %), and renewables (8 %). Energy consumption in the Northeast is expected to decline slightly over the next 25 years, though electricity consumption is projected to increase just slightly (Energy Information Administration 2012). Chapter 2 expands on this introduction to energy demand in the Northeast.

One notable aspect of energy consumption in the Northeast is the disproportionate reliance on residential fuel oil for home heating. In the Northeast fuel oil makes up about 13 % of the residential energy supply but nationally it is only about 3 % of the residential energy mix. The Northeast used nearly 0.4 EJ of wood for energy in 2010, which is about 16 % of the national total. This consumption has been relatively steady and reflects a slight decline from the peak of 0.6 EJ in 1986 (Energy Information Administration 2012). According to the U.S. Energy Information Administration (EIA) the cost of energy from wood has stayed relatively flat compared to increases in other energy sources (Fig. 1.6). Within the Northeast the wood price per gigajoule

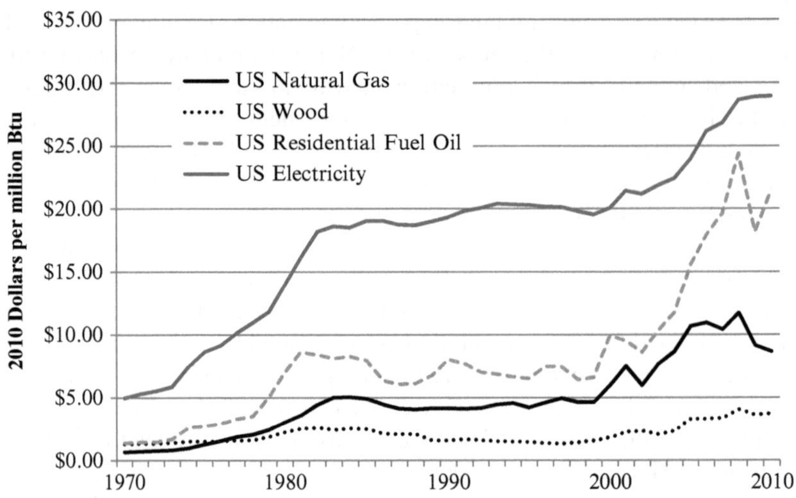

Fig. 1.6 Price of energy sources at the national level in 2010 dollars (EIA 2012)

reported by the EIA for 2010 ranges from about \$3 in Maine to over \$8 in West Virginia (Energy Information Administration 2012). In the Northeast, residential heating oil was between two and a half and nearly eight times more expensive per gigajoule in 2010 and averaged about \$21 per gigajoule. Of course the price of wood per Btu can be difficult to translate into a price paid to a landowner for a cord of wood. Local wood prices are specific to the form (i.e., cordwood, hogfuel, chips, pellets) and determined by a wide range of variables including distance to mills, availability of loggers, international markets, even weather conditions.

Current uses of wood for bioenergy in the Northeast range from personal wood stoves to industrial electricity generation. A considerable portion of the millions of cubic meters of fuel wood harvested in the Northeast is used to heat individual homes and, as of 2011, the Biomass Power Association listed 27 facilities in the Northeast that burn wood to generate up to 700 MW of electricity. However, one study suggests that even if all available woody biomass was used for electric generation in the northern United States it could only replace 19 % of coal-based electric generation (Goerndt et al. 2012). In between the individual and industrial scale bioenergy uses there are numerous intermediate scale users. For example, in 2010 Vermont had 43 schools that heated with wood. On average those schools used about 550 metric tons per year and generated about 4,474 gigajoules. Others producing bioenergy from wood include colleges, hospitals, prisons, and district heating systems such as the combined heat and power system in Concord, New Hampshire, that serves the local hospital. Existing bioenergy users of wood are already a significant addition to the demand generated by the traditional forest products industry, and should be part of supply assessments for new bioenergy facilities.

1.2.2 Wood Prices

At the national level softwood lumber prices have increased at a rate of about 5.7 % per year between 1965 and 1999 and then 2.0 % per year from 1999 to 2005 (Howard 2007). Hardwood lumber prices at the national level have increased at a rate of about 2.0 % per year in the United States since 1965. Prices in the Northeast have followed similar trends. Stumpage prices (the value of timber as it stands uncut) for roundwood (sawlogs and pulpwood) in Maine increased at an annual rate of 2.16 % between 1961 to 2000 and in Vermont, roundwood prices increased 3.35 % annually from 1982 to 2000 (Sendak et al. 2003). Red oak stumpage prices east of the Connecticut River increased from a median price of \$35 (2011 dollars) in 1978 to a median price of 75 (2011 dollars) per cubic meter in 2011 (Smith et al. 2012). Over this time, there were peaks when red oak prices were over \$212 (2011 dollars) per cubic meter in 1988, 1994, and 2001. It is worth highlighting the difference in prices across species. For example, the median white pine price was 5 % lower in 2011 than it had been in 1978 and the median hemlock price was nearly 60 % lower in 2011 (Smith et al. 2012). Pulpwood prices have been volatile in recent decades because of falling capacity and industry restructuring (Howard 2007). Pulpwood prices from 1965 to 2005 demonstrate not only the market volatility but

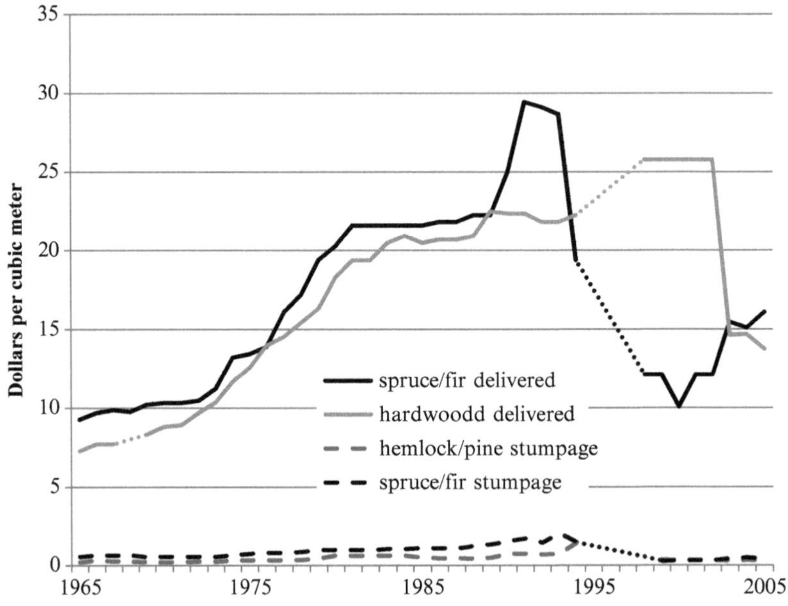

Fig. 1.7 Pulpwood prices in New Hampshire in 2005 dollars (Howard 2007)

also the huge difference between the stumpage price and the delivered price (Fig. 1.7). Stumpage prices were just 3 % of the delivered prices, reflecting the costs of harvesting and transporting wood.

1.2.3 Certification

One trend in forestry that can affect price and could influence use of wood for bioenergy is certification. In general, though some research has found certified wood can get a small (5 % or less) price premium, many companies have found access to some markets requires certification, particularly in low-grade markets such as pulp (Auld et al. 2008). For example, the European Commission may require some type of certification for pellets shipped from the United States to Europe to meet its 2020 renewable energy targets (Joudrey et al. 2012).

About 21 % of the timberland in the Northeast is certified compared with about 13 % nationwide (Lowe et al. 2011) (Fig. 1.8). In the Northeast about 4.2 million hectares are certified under Forest Stewardship Council (FSC), 3.3 million hectares are certified under Sustainable Forestry Initiative, (SFI), and 1.3 million hectares are certified under American Tree Farm System (ATFS). The FSC and SFI hectares include 1.9 million hectares that are dual certified. Maine has the largest amount and largest percentage of certified land almost at over 3 million hectares and 48 % of forestland. The influence of certification is growing. For example, between 2004 and 2009 both FSC and SFI doubled the number of certified hectares within their

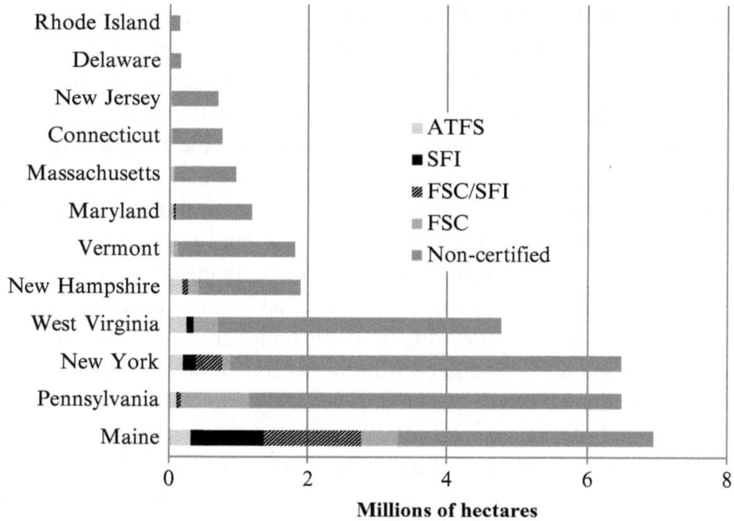

Fig. 1.8 Certified forests in the Northeast (Lowe et al. 2011)

respective programs (Fernholz et al. 2010). Certification is already pervasive enough in the northeast that most bioenergy projects could draw some material from certified lands, but currently it would be difficult to use only certified wood in a medium to large bioenergy project. The growth in certification and new group certifications for NIPFs could allow future bioenergy projects to focus entirely on certified forests.

The difference between certification systems from a bioenergy perspective is not entirely clear because all three systems are in the process of adapting to the bioenergy market. A recent review found that in general FSC certification caused companies to make more environmental or forest management changes while SFI certification precipitated more economic or system changes (Moore et al. 2012). While no certification system for the United States specifically addresses biomass harvests, certification can influence bioenergy harvests. For example, FSC certification requires that "management maintains, enhances, or restores habitat components and associated stand structures, in abundance and distribution that could be expected from naturally occurring processes"; these habitat components include "live trees with decay or declining health, snags, and well-distributed coarse down and dead woody material" (FSC 2009). This type of requirement could reduce the amount of biomass available from a harvest for bioenergy in order to ensure ecological sustainability.

1.2.4 Landowner Perspectives on Harvesting

Because noncorporate owners control the majority of forestland in the Northeast, private landowner perspectives on forest management and harvesting will have a substantial impact on the future of the bioenergy industry. The large urban population of the region has driven an emphasis on preservation and recreation for forest land accessible to the urban public (Irland 1999). Forests managed for timber and other products tend to be farther inland and farther from population centers. As mentioned above, population growth and movement has facilitated the fragmentation of forest landscapes and put pressure on landowners to subdivide and sell forests. Owners of small, fragmented pieces of forest are less likely to harvest timber (Wear et al. 1999). Survey results suggest that the top reason Northeast family forest owners keep their land is for its beauty, scenery, or privacy and timber production is much less important (Roper Public Affairs 2006, Butler 2008). The same survey indicated that in the Northeast the top reasons for harvesting timber is removal of trees damaged by natural catastrophe or tree maturity. Social constraints on forest harvesting can significantly reduce the availability of forest biomass. One study showed that social constraints (e.g., size of forest holding, harvest regulations, and owner attitudes) would reduce wood availability by an average of almost 60 % in the Northeast (Butler et al. 2010). These social constraints are greatest in small states such as Delaware, Rhode Island, and Maryland. In Massachusetts, NIPFs are likely to implement less intense harvests (i.e., thinning instead of regeneration harvests) than neighboring state lands (McDonald et al. 2006). A study in the Southeast found almost all consulting foresters' clients were willing to sell timber to energy facilities (Conrad et al. 2010). Whereas in Massachusetts a survey found the likelihood that family forest owners would include harvest biomass as part of a timber harvest was quite low (Markowski-Lindsay et al. 2012b). In fact, estimates of available forest biomass that include social constraints in Massachusetts are less than half of estimates that don't include those constraints (Markowski-Lindsay et al. 2012a). Bioenergy projects dependent on forest biomass from family forests will have to prioritize landowner education and outreach and emphasize the compatibility of biomass removal with maintaining the scenic beauty of forests. Chapter 14 discusses some of the other social issues surrounding bioenergy project development in the Northeast.

1.2.5 Production Capacity and Workforce

Another trend that influences how forests are managed in the Northeast is the decline in wood products manufacturing capacity and a decrease in the number of skilled woods workers. For example in 2007 there were 60 fewer mills operating in West Virginia than there had been in 2000 (Widmann et al. 2012). Between 2007 and 2009 the number of licensed loggers in West Virginia dropped by 40 % (Spong et al. 2010). Surveys in the Northeast suggest that many loggers are approaching retirement age and logging is perceived as a low-prestige career, both of which are leading to a decline in the number of loggers (Egan and Taggart 2004). Other factors that make logging a difficult business include the high fuel costs, insurance costs, equipment costs, and costs associated with workers compensation (Egan 2009).

1.3 Forestry in the Northeast

Forestry in the Northeast ranges from intensive management with clearcuts, herbicide applications, and relatively short rotations to single tree selection on small private ownerships. Some management regimes are designed to mimic natural disturbance, while others may have been designed to maximize economic return and many fall between these two extremes. In general partial harvesting is much more common than clearcutting, though there is considerable variation across the region (Canham et al. 2013). For example, Maine has higher harvest rates than southern New England and stands with low biomass (likely young stands) are more common than in southern New England (Canham et al. 2013). This is likely due to both the land ownership partner and the dominant forest type. Spruce-fir forests are often clearcut on a rotation of 30–60 years with a precommerical thinning and application of herbicides to control competing plants (Pitt and Lanteigne 2008, Bataineh et al. 2013). In northern hardwood forests even-aged management was more common in the 1970s and 1980s, but partial harvests have become more typical (Welsh and Healy 1993). As mentioned above, northern hardwood forests contain commercially important species (e.g. sugar maple, yellow birch) that are shade tolerant and can be managed through uneven-aged management. Northern hardwood stands are often managed on cutting cycles of 20 years (Leak et al. 1987). The greater light demands of oaks and hickories means that larger openings or removal of more trees is effective at encouraging regeneration in this forest type. Much of the success of regeneration harvests in oak-hickory forests is dependent on establishment of seedlings in the stand prior to harvest. Though there is tremendous variation in forest conditions, one estimate of rotation lengths for 51–61 cm sawtimber is between 75 and 90-years, assuming regular thinnings (Merritt 1980). Unfortunately, poor silvicultural practices are common in the region and only a small portion of family forests have a management plan in use (Munsell and Germain 2007, Butler 2008). A full discussion of the management of the regions forests is far beyond the scope of this chapter and better addressed by books on the subject (e.g., Barrett 1980b, Walker 1999).

Bioenergy has a role to play in many different types of forest management. Bioenergy can play an important role in providing new markets for forest products as the growth in the pellet market shows. In other cases, bioenergy could provide a market for low value wood that some landowners currently pay to remove in thinnings to improve the growth rates of the remaining trees. In some cases, biomass markets might encourage landowners to refrain from exploitive harvesting practices (Munsell and Germain 2007). Markets and prices determine which trees are considered sawtimber or pulpwood material and which are used for bioenergy. While it is unlikely that bioenergy projects will be able to pay landowners enough for bioenergy demand to compete with the demand for high quality sawtimber, there is likely to be competition with pulpwood and other traditional users where woodsheds overlap.

1.3.1 Potential Ecological Impact of Biomass Harvesting

Biomass harvesting can improve forest health through removal of invasive species, rehabilitation of degraded stands, or restoration of fire adapted ecosystems. There are also concerns about the potential for biomass harvesting to impinge on the forest's ability to support wildlife, provide clean water, sequester carbon, and regenerate a diverse suite of plants (Lattimore et al. 2009, Janowiak and Webster 2010). A potential effect of biomass harvests is a reduction in the quantity of dead wood left in the forest after harvest. Some biomass harvests focus on material such as tree tops, branches, or small trees that would be left as dead wood after timber harvests due to their low value. Dead wood (including downed woody material and snags) plays an important role in the ecosystem, from wildlife habitat to carbon storage (Evans and Kelty 2010, Evans 2011). Active forest management generally reduces the amount of dead wood in forests over the long term (Gore and Patterson 1986), although it often produces a short-term addition to dead wood (Liu et al. 2006). Reductions in dead wood impact wildlife because so many species rely on it for food or shelter (DeGraaf et al. 1992, Freedman et al. 1996). Part of the concern about reductions in dead wood is related to the nutrient stored in soils. Intensive cutting, such as clearcutting with whole-tree removal, can result in significant nutrient losses (Hornbeck et al. 1990). Forestry rules and regulations as well as guidelines specific to biomass harvests are in place to minimize the potential negative impacts of biomass harvesting.

1.3.2 Existing Forestry Best Management Practices and How They Relate to Biomass

The Northeast's forestry rules and regulations determine how forest biomass is harvested and how much biomass can be considered available for bioenergy. In addition, three Northeastern state governments have developed recommendations specific to biomass harvesting for bioenergy and these are addressed in the next section.

In the Northeast, best management practices (BMPs) have been established to protect water quality in forest operations and are determined at the state level. Programs vary from laws that prescribe mandatory practices to states that use voluntary BMPs and education and outreach programs. Vermont and New Hampshire rely on volunteer guidance, Massachusetts and Connecticut employ more regulations, and Rhode Island, New York, and Maine use a combination of approaches. BMPs usually include sections on timber harvesting, site preparation, reforestation, stream crossings, riparian management zones, prescribed burning and fire lines, road construction and maintenance, pesticides and fertilizers, and wetlands. These programs are routinely monitored and literature suggests that when these BMPs are properly implemented they do protect water quality (Shepard 2006). At the national level, the large majority of harvesting follows local BMPs. For example one study estimated that BMPs were used in 89 % of forestry operations (Ice et al. 2010).

In Maine, monitoring data showed that water quality BMPs were effectively applied in 77 % of stream crossings and 89 % of approaches to the crossings (MFS 2008).

Even though BMPs have traditionally focused on water quality, they can influence how and how much biomass is available for bioenergy projects. For example, in New Hampshire guidelines recommend leaving "some cull material" in the woods after a biomass harvest, although the specific amount is not defined (Chapman et al. 1997). The same guidelines provide recommendations related to soil productivity that would limit biomass removal on sites with nutrient-poor soils: specifically the use of bole-only harvesting (taking out the main portion of the tree stem, leaving branches and limbs in the woods) on low-fertility soils, or where fertility is unknown, as a precaution against nutrient loss. Other examples of ways BMPs can potentially limit biomass removals for bioenergy are through recommendations for the retention of snags, wildlife trees, downed logs, and riparian buffer zones (Evans et al. 2012).

Regulatory influences on bioenergy production are not limited to BMPs. The two wood-powered power plants in Vermont are required by the Public Service Board to ensure that their wood supply comes from sales with a harvest plan cleared by the Vermont Department of Forests, Parks and Recreation. New York has similar regulations but provides exemptions to properties that are accredited by FSC, SFI, or Tree Farm. Vermont also requires that whole-tree chip harvesters obtain a license (Title 10 V.S.A. Chapter 83, Section 2648). States such as Connecticut use certification of foresters and loggers as a way of encouraging professionalism and ecological protections.

While BMPs place limits on how forestry, including bioenergy harvests, is practiced, they may also expand opportunities for forestry and bioenergy. It has been argued that forestry requires a social license, in the sense that the general public desires input and influence on how forests are managed even on private lands (e.g., Spies et al. 2010). A good example of the social license for forestry comes from Maine where social pressure resulted in changes to the law to limit clearcuts to 101 ha (250 acres) or smaller. BMPs help reassure the public that forestry is being practiced in ways that protect public amenities such as clean water and scenic views. Without these assurances, public opposition to harvests would likely be greater. Social acceptance is one of the driving forces behind the creation of new retention and harvesting guidelines specifically for biomass removals for the bioenergy market.

1.3.3 Biomass Harvesting Guidelines

A number of states in the region including Maine, Maryland, and Pennsylvania have created guidelines specific to biomass harvests. These biomass harvesting guidelines are designed to fill the gaps where existing BMPs may not be sufficient to protect forest resources under new biomass harvesting regimes. Existing guidelines did not anticipate the increased rate or new methods of biomass removal and offer no specific guidance on the amount of extraction that is acceptable for meeting a range of forest management objectives. For example, Pennsylvania's old BMPs encouraged operators "to use as much of the harvested wood as possible to minimize debris," while the new guidelines recommend leaving "15–30 % of harvestable biomass as coarse woody debris" (PA DCNR 2008, Bauer and Weisberg 2009). Intensification of biomass utilization, particularly for energy and fuel needs, presents a range of potential environmental risks (Lattimore et al. 2009, Janowiak and Webster 2010). As with certification, new management regulations can increase forest management costs (Van Deusen et al. 2012), but these costs may be offset by access to new markets. Similarly, certification of new management regulations may increase social acceptance of harvesting thereby reducing legal costs or social opposition.

Pennsylvania's guidelines are a direct result of increased interest in woody biomass for energy. The passage of Pennsylvania's Alternative Energy Portfolio Standards Act (Act 213 of 2004) helped drive that interest by requiring "all load-serving energy companies in the state to provide 18 % of their electricity using alternative sources by the year 2020." Pennsylvania's guidelines are unusual in that they include comments on biomass policy and a supply assessment (PA DCNR 2008). For example, the guidelines suggest that facilities requiring 2,000 t per year are better suited to Pennsylvania than larger facilities. Since Pennsylvania's state forestlands are FSC certified, their biomass harvesting guidelines directly reference FSC standards. Pennsylvania's DCNR uses the FSC's Appalachian Regional Standards, but the state biomass harvesting guidelines provide greater specificity on woody biomass removals. For example, the FSC standard requires that "measures to protect streams from degradation of water quality and/or their associated aquatic habitat are used in all operations." The Pennsylvania biomass guidelines extend this idea by adding "biomass harvesting of any materials along stream and river banks or along bodies of water is unacceptable."

In Maine, "guidelines specific to woody biomass retention (were) missing from existing best management practices and regulations." (MFS et al. 2008) Therefore, the state undertook a collaborative effort between the Maine Forest Service, the University of Maine, and the Trust to Conserve Northeast Forestlands to develop woody biomass retention guidelines. The resulting recommendations lack specific targets and provide flexibility for the forester or landowner. For example, they suggest forest litter should be left on-site "to the extent possible" and that operators should "minimize removal" of fine woody material on low-fertility sites and leave "as much fine woody material as possible" without the specific guidelines for downed woody material retention found in some other state guidelines (Benjamin 2010).

Maryland created a set of biomass harvesting and retention guidelines in fall of 2010 through a collaborative process that included state agencies, universities, and nongovernmental organizations (Pinchot Institute and MDNR 2010). The guidelines focus on soil risks, wildlife habitat, water quality, and silviculture. Notably the guidelines include a map that uses physical and chemical properties of soils to identify areas at higher risk of negative impacts from biomass removal. Maryland's guidelines also differentiate between mixed-hardwood forests and softwood dominant plantations.

In general, biomass harvesting and retention guidelines in the Northeast address dead wood, wildlife, water quality, riparian zones, soil productivity, silviculture, and disturbance. Most guidelines include recommendations for the retention of slash or dead wood post-harvest that could constrain the supply of biomass for bioenergy. For instance, Pennsylvania's guidelines suggest leaving 15–30 % of harvestable biomass on site to protect soil fertility and wildlife habitat. Similarly, Maine and Pennsylvania both suggest leaving all snags possible (Evans et al. 2010). Other states in the region are considering the creation of specific biomass guidelines or incorporation of biomass harvesting in existing regulatory or BMP frameworks. For example, Vermont's Study Committee on Biomass Energy Development recommended the legislature create wood procurement standards that include harvest guidelines. Based on the expansion of regulatory or voluntary guidance for forest biomass harvest in the Northeast, bioenergy project developers should plan to integrate harvest and retention guidelines into supply assessments and procurement strategies.

1.4 Future of Forests, Forestry, and Bioenergy in the Northeast

This chapter provides a snapshot of the current forest and forestry context for bioenergy in the Northeast, but of course these systems are dynamic and bioenergy project planners must prepare for change. Two of the most significant changes that could affect future wood supply are a warming climate and shifting precipitation regime. Summers in the Northeast could be 3.3–7.8 °C warmer and winters could be 4.4–6.7 °C warmer than historic averages by 2100 (Frumhoff et al. 2007). Winter precipitation may increase 11–14 % over the century with a greater proportion falling as rain rather than snow (Huntington et al. 2004, Hayhoe et al. 2007). A warming climate may cause snow to melt earlier in the year and therefore decrease sublimation and produce an earlier and larger peak runoff (Dankers and Christensen 2005, Hayhoe et al. 2007). The warming climate is likely to increase the importance of oak-hickory forests at the expense of spruce-fir and northern hardwood forests (Iverson et al. 2008, Tang and Beckage 2010). Unfortunately a whole suite of invasive pests threaten to exacerbate the negative impacts of climate including hemlock woolly adelgid (Orwig et al. 2012), emerald ash borer (Pugh et al. 2011), and Asian longhorned beetle (Dodds and Orwig 2011). A changing climate may even bring aggressive subtropical exotic plant species such as kudzu (*Pueraria Montana*) to Northeastern forests (Jarnevich and Stohlgren 2009). In response to the negative impacts of climate change and invasive species managers and scientists are developing new forest management strategies (Millar et al. 2007, Evans and Perschel 2009, Likens and Franklin 2009). Bioenergy can be part of the solution because demand for low-grade wood for energy can provide forest managers new opportunities. For example, a bioenergy harvest could include the salvage removal of hemlock trees killed by woolly adelgid and facilitate site preparation for planting or restoration.

Landownership and the forestry industry in the Northeast are also changing. Currently, the forestland base provides ample opportunities for harvesting forest biomass for energy. However, that may not always be true. Fragmentation of forest land through further spread of ex-urbanization development and subdivision of large parcels could reduce the traditional timber industry and also constrain the development of a bioenergy industry based on wood.

Along with the threat of a shrinking timber industry, workforce limitations could hamper bioenergy projects. Without skilled foresters, loggers, and truckers to select, cut, and transport wood it would be impossible to develop a sustainable bioenergy industry. Programs to train and support those working in the forest products industry could help encourage a new generation to pursue careers in woods work. Increased mechanization, a trend across the forest products industry, could also help increase interest in forest industry jobs by making the work less strenuous and less dangerous. The development of a bioenergy industry would augment the trend towards greater mechanization because bioenergy harvests tend to focus on small diameter trees which are more efficient to remove with a mechanized harvester or feller-buncher.

While government regulations and incentive programs change too rapidly to provide a detailed current account in this volume, it is worth noting their potential impact on bioenergy projects. The increasing demand for wood pellets in Europe indicates that inclusion of biomass in renewable energy targets or renewable portfolio standards provides a strong incentive to expand the bioenergy sector. On the other hand, regulations can stifle bioenergy project development by excluding certain type of biomass utilization or bioenergy generation. Recent rule changes in Massachusetts exclude plants with efficiency rates of less than 50 % which will exclude most electricity-generating facilities currently operating and present regulatory obstacles for new plants. Carbon accounting was at the heart of the rule change in Massachusetts. At the time of this writing, the U.S. Environmental Protection Agency is set to make a decision on the carbon impact of bioenergy which could either encourage Massachusetts to change is renewable energy credit rules or encourage other states in the Northeast to follow Massachusetts's path. The most challenging aspect of regulations for the bioenergy industry can be adapting to their rapid changes. For example, the abrupt changes to the biomass crop assistance program, which provided matching payments from the federal government for the harvesting and transportation of biomass materials, caused difficulties for both biomass producers who budgeted on the subsidies and consumers who planned on the expanded supply. Bioenergy project developers in the Northeast will need to continue to watch the regulatory environment to ensure they are prepared for new requirements or opportunities.

Bioenergy projects based on the Northeast's abundant wood resources will be developed within the environmental context of forests that vary significantly across relatively small distances. Similarly, landownership patterns, forest industry capacities, and regulations differ throughout the region. The addition of a changing climate, forest fragmentation trends, and shifts in the workforce emphasize the importance of understanding the forest and forestry context for those implementing bioenergy projects based on wood

References

Anderson NM, Germain RH, Bevilacqua E (2011) Geographic information system-based spatial analysis of sawmill wood procurement. J Forest 109(1):34–42

Auld G, Gulbrandsen LH, McDermott CL (2008) Certification schemes and the impacts on forests and forestry. Annu Rev Environ Resour 33(1):187–211

Barrett JW (1980a) The northeastern region. In: Barrett JW (ed) Regional silviculture of the United States. Wiley, New York, NY, pp 25–65

Barrett JW (ed) (1980b) Regional silviculture of the United States. Wiley, New York, NY

Bataineh MM, Wagner RG, Weiskittel AR (2013) Long-term response of spruce–fir stands to herbicide and precommercial thinning: observed and projected growth, yield, and financial returns in central Maine, USA. Can J Forest Res 43(4):385–395

Bauer JM, Weisberg PJ (2009) Fire history of a central Nevada Pinyon-Juniper woodland. Can J Forest Res 39(8):1589–1599

Benjamin JG (2010) Considerations and recommendations for retaining woody biomass on timber harvest sites in Maine. Miscellaneous publication 761, University of Maine, Maine Agricultural and Forest Experiment Station, Orono, ME

Brose PH, Waldrop TA (2006) Fire and the origin of table mountain pine-pitch pine communities in the southern Appalachian Mountains, USA. Can J Forest Res 36(3):710–718

Butler BJ (2008) Family forest owners of the United States, 2006. GTR-NRS-27, USDA Forest Service, Northern Research Station, Newtown Square, PA

Butler BJ, Ma Z, Kittredge DB, Catanzaro P (2010) Social versus biophysical availability of wood in the northern United States. North J Appl For 27:151–159

Canham CD, Rogers N, Buchholz T (2013) Regional variation in forest harvest regimes in the northeastern United States. Ecol Appl 23(3):515–522

Chapman C, Donovan G, Fay S, Foss C, Jones G et al (1997) Good forestry in the granite state: recommended voluntary forest management practices for New Hampshire. New Hampshire Division of Forests & Lands and Society for the Protection of New Hampshire Forests, Concord, NH

Cogbill CV, Burk J, Motzkin G (2002) The forests of presettlement New England, USA: spatial and compositional patterns based on town proprietor surveys. J Biogeogr 29(10–11): 1279–1304

Conrad JLI, Bolding MC, Aust WM, Smith RL (2010) Wood-to-energy expansion, forest ownership changes, and mill closure: consequences for U.S. South's wood supply chain. Forest Policy Econ 12(6):399–406

Daigle JJ, Utley L, Chase LC, Kuentzel WF, Brown TL (2012) Does new large private landownership and their management priorities influence public access in the northern forest? J Forest 110(2):89–96

Dankers R, Christensen OB (2005) Climate change impact on snow coverage, evaporation and river discharge in the Sub-Arctic Tana basin, northern Fennoscandia. Clim Chang 69(2):367–392

DeGraaf RM, Yamasaki M, Leak WB, Lanier JW (1992) New England wildlife: management forested habitats. General technical report NE-144, USDA Forest Service, Northeastern Forest Experiment Station, Radnor, PA

Dodds KJ, Orwig DA (2011) An invasive urban forest pest invades natural environments – Asian longhorned beetle in northeastern US hardwood forests. Can J Forest Res 41(9):1729–1742

Drummond MA, Loveland TR (2010) Land-use pressure and a transition to forest-cover loss in the eastern United States. Bioscience 60(4):286–298

Easterling DR (2002) Recent changes in frost days and the frost-free season in the United States. Bull Am Meteorol Soc 83(9):1327–1332

Egan A (2009) Characteristics of New York's logging businesses and logging business owners. North J Appl For 26(3):106–110

Egan A, Taggart D (2004) Who will log in Maines north woods? A cross-cultural study of occupational choice and prestige. North J Appl For 21(4):200–208

Egan A, Taggart D, Annis I (2007) Effects of population pressures on wood procurement and logging opportunities in northern New England. North J Appl For 24(2):85–90
Energy Information Administration (2012) Annual energy outlook. http://www.eia.gov/forecasts/aeo/. Accessed 20 August 2012. Washington, DC
Evans AM (2011) Ecology of dead wood in the southeast. Forest Guild, Santa Fe, NM
Evans AM, Kelty MJ (2010) The ecology of dead wood in the Northeast. Forest Guild, Santa Fe, NM
Evans AM, Perschel RT (2009) A review of forestry mitigation and adaptation strategies in the northeast U.S. Clim Chang 96(1–2):167–183
Evans AM, Perschel RT, Kittler BA (2010) Revised assessment of biomass harvesting and retention guidelines. The Forest Guild, Santa Fe, NM
Evans AM, Perschel RT, Kittler BA (2012) Overview of forest biomass harvesting guidelines. J Sustain For 32(1–2):89–107
Eyre FH (1980) Forest cover types of the United States and Canada. Society of American Foresters, Bethesda, MD
Fernholz K, Howe J, Bratkovich S, Bowyer J (2010) Forest certification: a status report. Dovetail Partners, Minneapolis, MN
Freedman B, Zelazny V, Beaudette D, Fleming T, Johnson G et al (1996) Biodiversity Implications of changes in the quantity of dead organic matter in managed forests. Environ Rev 4(3):238–265
Frumhoff PC, McCarthy JJ, Melillo JM, Moser SC, Wuebbles DJ (2007) Confronting climate change in the U.S northeast: science, impacts and solutions. Northeast climate impacts assessment. Union of Concerned Scientists, Cambridge, MA
FSC (2009) US forest management standard draft 8.1. Forest Stewardship Council, Washington, DC
Gavier-Pizarro GI, Radeloff VC, Stewart SI, Huebner CD, Keuler NS (2010) Housing is positively associated with invasive exotic plant species richness in New England, USA. Ecol Appl 20(7): 1913–1925
Goerndt ME, Aguilar FX, Miles P, Shifley S, Song N et al (2012) Regional assessment of woody biomass physical availability as an energy feedstock for combined combustion in the US northern region. J Forest 110(3):138–148
Gore JA, Patterson WA III (1986) Mass of downed wood in northern hardwood forests in New Hampshire: potential effects of forest management. Can J Forest Res 16(2):335–339
Hammer RB, Stewart SI, Radeloff VC (2009) Demographic trends, the wildland–urban interface, and wildfire management. Soc Nat Resour 22(8):777–782
Hayhoe K, Wake C, Anderson B, Liang X-Z, Maurer E et al (2008) Regional climate change projections for the northeast USA. Mitig Adapt Strateg Glob Chang 13(5–6):425–436
Hayhoe K, Wake CP, Huntington TG, Luo L, Schwartz MD et al (2007) Past and future changes in climate and hydrological indicators in the US northeast. Clim Dyn 28(4):381–407
Hornbeck JW, Smith CT, Martin QW, Tritton LM, Pierce RS (1990) Effects of intensive harvesting on nutrient capitals of three forest types in New England. For Ecol Manag 30(1–4):55–64
Howard JL (2007) U.S. timber production, trade, consumption, and price statistics 1965 to 2005. FPL-RP-637, Forest Service, Forest Products Laboratory, Madison, WI
Huntington TG, Hodgkins GA, Keim BD, Dudley RW (2004) Changes in the proportion of precipitation occurring as snow in New England (1949–2000). J Clim 17(13):2626–2636
Ice GG, Schilling EG, Vowell JG (2010) Trends for forestry best management practices implementation. J Forest 108(6):267–273
Irland LC (1999) The Northeast's changing forests. Harvard Forest, Petersham, MA
Iverson L, Prasad A, Matthews S (2008) Modeling potential climate change impacts on the trees of the northeastern United States. Mitig Adapt Strateg Glob Chang 13(5–6):487–516
Janowiak MK, Webster CR (2010) Promoting ecological sustainability in woody biomass harvesting. J Forest 108(1):16–23
Jarnevich C, Stohlgren T (2009) Near term climate projections for invasive species distributions. Biol Invasions 11(6):1373–1379

Jordan MJ, Patterson WA III, Windisch AG (2003) Conceptual Ecological models for the Long Island pitch pine barrens: implications for managing rare plant communities. For Ecol Manag 185(1–2):151–168

Joudrey J, McDow W, Smith T, Larson B (2012) European power from U.S. forests. Environmental Defense Fund, New York, NY

Lattimore B, Smith CT, Titus BD, Stupak I, Egnell G (2009) Environmental factors in woodfuel production: opportunities, risks, and criteria and indicators for sustainable practices. Biomass Bioenergy 33(10):1321–1342

Leak WB, Solomon DS, DeBald PS (1987) Silvicultural guide for northern hardwood types in the northeast (revised). RP-NE-603, USDA Forest Service, Northeastern Forest Experiment Station, Broomall, PA

Likens GE, Franklin JF (2009) Ecosystem thinking in the northern forest and beyond. Bioscience 59(6):511–513

Liu W, Bryant D, Hutyra L, Saleska S, Hammond-Pyle E et al (2006) Woody debris contribution to the carbon budget of selectively logged and maturing mid-latitude forests. Oecologia 148(1):108–117

Londré RA, Schnitzer SA (2006) The distribution of lianas and their change in abundance in temperate forests over the past 45 years. Ecology 87(12):2973–2978

Loucks C, Olson D, Dinerstein E, Weakley A, Noss R et al (1999) Appalachian mixed mesophytic forests. In: Ricketts TH, Carney K, Abell R, Walters S (eds) Terrestrial ecoregions of north America: a conservation assessment ecoregions assessments. Island, Washington, DC

Lowe L, Brogan S, McClure N, Nowak J, Oates B et al (2011) Forest certification programs: status and recommendations in the south. Southern Group of State Foresters, Winder, GA

Luppold W, Mace T, Weatherspoon A, Jacobson K (2011) Changes in the fuel pellet industry in the Lake States, 2005 to 2008. North J Appl For 28(4):204–207

Markowski-Lindsay M, Catanzaro P, Damery D, Kittredge DB, Butler BJ et al (2012a) Forest-based biomass supply in Massachusetts: how much is there and how much is available. J Environ Manag 106:1–7

Markowski-Lindsay M, Stevens T, Kittredge DB, Butler BJ, Catanzaro P et al (2012b) Family forest owner preferences for biomass harvesting in Massachusetts. Forest Policy Econ 14(1):127–135

McCaskill GL, McWilliams WH, Barnett CJ, Butler BJ, Hatfield MA et al (2011) Maine's forests 2008. USDA Forest Service, Northern Research Station, Newtown Square, PA

McDonald RI, Motzkin G, Bank MS, Kittredge DB, Burk J et al (2006) Forest harvesting and land-use conversion over two decades in Massachusetts. For Ecol Manag 227(1–2):31–41

McNab WH, Cleland DT, Freeouf JA, Keys JE Jr, Nowacki GJ et al (2007) Description of ecological subregions: sections of the conterminous United States. GTR-WO-76B, U.S. Forest Service, Washington, DC

Merritt C (1980) The Central Region. In: Barrett JW (ed) Regional silviculture of the United States. Wiley, New York, NY, pp 107–144

MFS (2008) Maine forestry best management practices use and effectiveness 2006–2007. Maine Forest Service, Augusta, ME

MFS, University of ME, and TCNF (2008) Biomass retention guidelines for timber harvesting in Maine. Version 4. Maine Forest Service, University of Maine, and the Trust to Conserve Northeast Forestlands, Orono, ME

Millar C, Stephenson N, Stephens S (2007) Climate change and forests of the future: managing in the face of uncertainty. Ecol Appl 17(8):2145–2151

Moore SE, Cubbage F, Eicheldinger C (2012) Impacts of forest stewardship council (FSC) and sustainable forestry initiative (SFI) forest certification in north America. J Forest 110(2):79–88

Munsell JF, Germain RH (2007) Woody biomass energy: an opportunity for silviculture on nonindustrial private forestlands in New York. J Forest 105:398–402

National Climate Data Center (2012) US climate at a glance. http://www.ncdc.noaa.gov/oa/climate/research/cag3/cag3.html. Accessed 8 July 2012

Orwig DA, Thompson JR, Povak NA, Manner M, Niebyl D et al (2012) A foundation tree at the precipice: tsuga canadensis health after the arrival of adelges tsugae in central New England. Ecosphere 3(1):art10

PA DCNR (2008) Guidance on harvesting woody biomass for energy. Pennsylvania Department of Conservation and Natural Resources, Harrisburg, PA

Pinchot Institute, and MDNR (2010) A guide to forest biomass harvesting and retention in Maryland. Pinchot Institute for Conversation and Maryland Department of Natural Resources, Forest Service, Annapolis, MD

Pitt D, Lanteigne L (2008) Long-term outcome of precommercial thinning in northwestern New Brunswick: growth and yield of balsam fir and red spruce. Can J Forest Res 38(3):592–610

Pugh SA, Liebhold AM, Morin RS (2011) Changes in ash tree demography associated with emerald ash borer invasion, indicated by regional forest inventory data from the Great Lakes states. Can J Forest Res 41(11):2165–2175

Radeloff VC, Nelson E, Plantinga AJ, Lewis DJ, Helmers D et al (2011) Economic-based projections of future land use in the conterminous United States under alternative policy scenarios. Ecol Appl 22(3):1036–1049

Roper Public Affairs (2006) Family forest owners: an in-depth profile. The sustaining family forests initiative. Yale School of Forestry and Environmental Studies, New Haven, CT

Sendak PE, Abt RC, Turner RJ (2003) Timber supply projections for northern New England and New York: integrating a market perspective. North J Appl For 20(4):175–185

Shepard JP (2006) Water quality protection in bioenergy production: the US system of forestry best management practices. Biomass Bioenergy 30(4):378–384

Shifley SR, Aguilar FX, Song N, Stewart SI, Nowak DJ et al (2012) Forests of the northern United States. GTR-NRS-90, USDA Forest Service, Northern Research Station, Newtown Square, PA

Smith JS, Markowski-Lindsay M, Wagner JE, Kittredge DB (2012) Stumpage prices in southern New England (1978–2011): how do red oak, white pine, and hemlock prices vary over time? North J Appl For 29(2):97–101

Smith WB, Miles PD, Perry CH, Pugh SA (2009). Forest resources of the United States, 2007. GTR-WO-78, US Forest Service, Washington, DC

Spelter H, Toth D (2009) North America's wood pellet sector. FPL-RP-656, USDA Forest Service, Forest Products Laboratory, Madison, WI

Spies T, Giesen T, Swanson F, Franklin J, Lach D et al (2010) Climate change adaptation strategies for federal forests of the Pacific northwest, USA: ecological, policy, and socio-economic perspectives. Landsc Ecol 25(8):1185–1199

Spong B, Wang J, Summerfield D (2010) Characteristics of west Virginia loggers during economically difficult times. (pp 579–585). In: Fei S, Lhotka JM, Stringer JW, Gottschalk KW, Miller GW (eds) 17th central hardwood forest conference. GTR-NRS-P-78, USDA Forest Service, Lexington, KY

Stein SM, Carr MA, McRoberts RE, Mahal LG, Comas SJ (2010). Threats to at-risk species in America's private forests: a forests on the edge report. GTR-NRS-73, USDA Forest Service, Northern Research Station, Newtown Square, PA

Stone RS, Tyrrell ML (2012) Motivations for family forestland parcelization in the Catskill/Delaware watersheds of New York. J Forest 110(5):267–274

Tang G, Beckage B (2010) Projecting the distribution of forests in New England in response to climate change. Divers Distrib 16(1):144–158

Theobald D (2010) Estimating natural landscape changes from 1992 to 2030 in the conterminous US. Landsc Ecol 25(7):999–1011

US Census Bureau (2010) Decennial census, summary file 1–4. US Census Bureau, Washington, DC

USFS (2012) State and private forestry annual report: appendix. National Association of State Foresters and the USDA Forest Service, Washington, DC

Van Deusen P, Wigley TB, Lucier AA (2012) Cumulative effects of constraints on forest management. J Forest 110(3):123–128

Walker LC (1999) The north american forests. CRC Press, Boca Raton, FL

Wear DN, Liu R, Michael Foreman J, Sheffield RM (1999) The effects of population growth on timber management and inventories in Virginia. For Ecol Manag 118(1–3):107–115
Weinberg A (2012) Retaining working forests: eastern north Carolina. Open Space Institute, New York, NY
Welsh CJ, Healy WM (1993) Effect of even-aged timber management on bird species diversity and composition in northern hardwoods of New Hampshire. Wildl Soc Bull 21(2):143–154
Westveld M (1956) Natural forest vegetation zones of New England. J Forest 54(5):332–338
Widmann RH, Cook GW, Barnett CJ, Butler BJ, Griffith DM et al (2012). West Virginia's forests 2008. RB-NRS-61, USDA Forest Service, Northern Research Station, Newtown Square, PA

Chapter 2
The Economic Availability of Woody Biomass Feedstocks in the Northeast

Matthew Langholtz and Michael Jacobson

2.1 Introduction

"Future growth of the U.S. bioenergy industry will depend on the cost, quality, and quantity of biomass available to biorefineries" (USDOE 2012). Biomass supply assessments aim to quantify and characterize biomass supplies to inform investment decisions, research and development efforts, and policy development. Biomass supply evaluations range from localized, project-specific assessments to regional or national estimates. Assessing feedstock supplies for bioenergy is not unlike approaches applied in agricultural or forest sectors. Supply evaluation requires assumptions of what product specifications (e.g., species, ash content, particle size) can be used for a given commercial process, what geographic area is to be considered, and what land types can be accessed for production. The long-term environmental, economic, and social sustainability of feedstock supplies must also be considered. Evaluating the economic availability of feedstocks requires consideration of costs of production and competition from alternative markets. Typically, more (and more desirable) feedstock supplies become available at higher prices. Thus, to evaluate the economic availability of feedstocks, it is useful to evaluate how feedstock supplies change with price or, alternatively, what feedstock price might be needed to incentivize production to meet an anticipated level of demand.

One source of increased demand for biomass in the future is expected to be for cellulosic feedstocks for the production of biofuels, biopower, and bioproducts. To enhance the US energy security, reduce dependence on oil, provide

M. Langholtz, Ph.D. (✉)
Environmental Sciences Division, Oak Ridge National Laboratory,
P.O. Box 2008, 1 Bethel Valley Road, Oak Ridge, TN 37831-6335, USA
e-mail: langholtzmh@ornl.gov

M. Jacobson
Department of Ecosystem Science and Management, Pennsylvania State University,
Forest Resources Building 309, 16802 University Park, Pennsylvania, USA

M. Jacobson and D. Ciolkosz (eds.), *Wood-Based Energy in the Northern Forests*,
DOI 10.1007/978-1-4614-9478-2_2, © Springer Science+Business Media New York 2013

environmental benefits, and create economic opportunities, the Energy Independence and Security Act of 2007 (EISA) aims to support the production and use of 136 billion liters (36 billion gallons) of renewable fuels by 2022. This ramp-up of biofuels use includes second-generation cellulosic biofuels and biomass-based diesel and must accommodate additional projected demand for biopower (i.e., electricity generation from biomass). Meeting this combined demand for biofuels and biopower is expected to require about 295 million dry Mg of biomass by the year 2022 (Langholtz et al. 2012). With a large population and a significant forestry and agricultural base, the Northeast is well positioned to be both a consumer and a producer of bioenergy. Considering this potential future production, here we evaluate the projected economic availability of woody biomass feedstocks in 12 Northeast US states: Connecticut, Delaware, Maine, Maryland, Massachusetts, New Hampshire, New Jersey, New York, Pennsylvania, Rhode Island, Vermont, and West Virginia.

2.2 Biomass Resource Assessments

In 2005, DOE released "Biomass as Feedstock for a Bioenergy and Bioproducts Industry: The Technical Feasibility of a Billion-Ton Annual Supply" known as the "Billion-Ton Study." The Billion-Ton Study was an estimate of potential biomass supply in the United States based on assumptions about inventory, production capacity, and technology (Perlack et al. 2005). Results from the study suggest that US agriculture and forest resources have the capability to sustainably produce a billion dry tons of biomass annually—enough to displace approximately 30 % of the country's present petroleum consumption. While significant as a first broad national assessment of the quantity of biomass that might be available, this report did not address the question of what price might be needed to procure this biomass—bearing in mind that more supply becomes available as the price offered for the biomass is increased.

As a follow-up to the 2005 BTS, in 2011 the DOE released "U.S. Billion-Ton Update: Biomass Supply for a Bioenergy and Bioproducts Industry" or "BT2" (U.S. Department of Energy 2011), which evaluates the economic availability of biomass feedstocks in the USA. Applying a macroeconomic simulation model of the agricultural sector (see De la Torre Ugarte and Ray 2000), the BT2 projects biomass feedstock supplies and prices between 2012 and 2030 at the county level (results available at www.bioenergykdf.net), coincident with implementation of EISA and DOE initiatives. The BT2 includes projections of feedstock supplies available at multiple price points and places greater emphasis on environmental sustainability than the 2005 study. Findings include the following: (1) there is enough resource potential to meet the 2022 advanced biofuel goals assuming a farmgate price of about $50 per dry ton, (2) potential resources are widely distributed, and (3) energy crops are the single largest source of new feedstock potentially available. While others have developed national biomass resource assessments with generally comparable results (e.g., Khanna et al. 2011) none to date have incorporated both agricultural and forest resources with economic projections at the county level.

Biomass supply assessments have been released for some states in the Northeast. For example, it has been estimated that between 3.9 and 7.6 million dry Mg of forest biomass is potentially available annually in the state of New York (Wojnar 2012), and between 68,000 and 401,000 Mg of forest biomass is potentially available in Massachusetts, depending on price (Walker et al. 2010). To assess potential woody biomass feedstock supplies in the 12 Northeast states, in this chapter we evaluate supply and price projections of the 12 Northeast states as projected in the Billion-Ton Update (U.S. Department of Energy 2011) and make comparisons with these state-level assessments.

2.3 Billion-Ton Update Modeling Assumptions

The economic availability of biomass feedstocks in the BT2 is assessed using POLYSYS, a macroeconomic simulation model of the USA (De la Torre Ugarte and Ray 2000; U.S. Department of Energy 2011). Woody biomass resources are modeled in one of the two categories: (1) short-rotation woody crops (SRWCs), which are simulated to compete economically with other crops (both conventional crops and dedicated biomass feedstocks) on agricultural land, and (2) forest resources, such as thinnings, logging residues, and some pulpwood, from forest land. In this land-use allocation modeling framework it is assumed that SRWCs and other dedicated crops are not established on forestland, as no forestland conversion is simulated. Here we describe the POLYSYS modeling framework in general, and then focus on assumptions used in the woody biomass feedstocks in the Northeast USA.

2.3.1 Modeling Framework

The POLYSYS modeling framework can be conceptualized as a variant of an equilibrium displacement model (EDM). EDMs establish simultaneous systems of generalized functions where endogenous variables (e.g., quantities and prices) are measured as proportionate changes and are a function of proportionate changes in exogenous, or curve-shifting, variables (e.g., costs of production and available resources). The equilibrium market is shocked exogenously (e.g., demand for production is changed), and the impacts of this disturbance are approximated by linear combinations of the products of the exogenous variables and their elasticities.

The wide appeal of EDMs in analytical work results in part from their flexibility in modeling a wide variety of market structures. The simplicity of an EDM also contributes to its usefulness and popularity (Aliston et al. 1995; Piggot 1992). Given the elasticities associated with prices and curve-shifter variables, estimates of the impacts associated with a percentage change in an exogenous variable (or multiple variables) may be easily and quickly obtained.

POLYSYS was developed to simulate changes in economic policy, agricultural management, and natural resource conditions and to estimate the resulting impacts

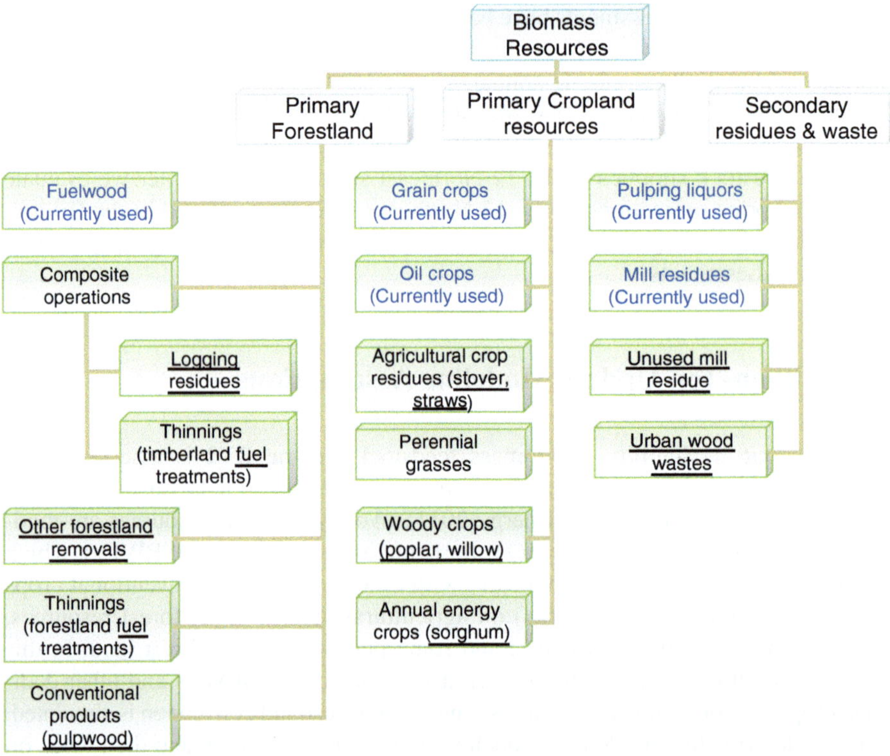

Fig. 2.1 Taxonomy of terrestrial biomass feedstocks. *Underlined* feedstocks are categories reported in this study. Sorghum and switchgrass are combined and reported as "herbaceous energy crops"; poplars and willows are combined and reported as "woody energy crops"

from these changes on the US agricultural sector (De la Torre Ugarte and Ray 2000; Lin et al. 2000; Ray et al. 1998). At its core, POLYSYS is structured as a system of interdependent modules simulating (1) crop supply for the continental USA, which is disaggregated into 3,110 production regions; (2) national crop demands and prices; (3) national livestock supply and demand; and (4) agricultural income. Variables that drive the modules include planted and harvested area, production inputs, yield, exports, costs of production, demand by use, commodity price, government program outlays, and net realized income. Conventional crops currently considered in POLYSYS include corn, grain sorghum, oats, barley, wheat, soybeans, cotton, rice, and hay. POLYSYS also considers terrestrial biomass crops for bioenergy. These crops include annual and perennial herbaceous energy crops and coppice and non-coppice woody crops for bioenergy and the collection of crop and woody residues as additional cellulosic feedstocks. The taxonomy of these modeled feedstocks is illustrated in Fig. 2.1.

The regional crop supply module in POLYSYS consists of 3,110 independent linear programming regional models that correspond to county boundaries. Each county is characterized by homogeneous production for all cropland area by crop type and tillage. The purpose of the crop supply module is to allocate land at the county level to the model crops given baseline information on county land area of the model crops, regional enterprise budgets of each crop, prices from the previous year, and a set of allocation rules. County-level acreages of various crops are provided assuming a 3-year average of observed cropland production by NASS, which encompass 126 million hectares (311 million acres) of national cropland.

For estimates of available pastureland in farms, the 2007 Census of Agriculture was queried against certain exclusion criteria related to best management practices of energy crop establishment and sustainability considerations. In the 2007 Census, 169 million hectares (417 million acres) were classified as non-wooded pastureland (using the specific classifications "Pastureland" and "Cropland used as Pasture"). It is assumed that pastureland capable of transitioning to dedicated terrestrial biomass production is restricted within POLYSYS to counties which lie east of the 100th meridian that are neither irrigated nor wooded. This exclusion is the result of insufficient evidence that terrestrial feedstocks can be successfully established on pastureland without supplemental or permanent irrigation. Since the irrigated portion of cropland or cropland pasture is reported as a separate category "irrigated pasture," we identify nonirrigated pasture by assuming that irrigation occurs at equal rates across both pasture and cropland pasture at the county level.

Irrigated pasture acreage exceeding combined nonirrigated pastureland and cropland pasture occurs in 48 counties, with all but 9 of those counties located west of the 100th meridian. This amount of irrigated acreage exceeding permanent pasture and cropland pasture is significant, totaling 1.25 million hectares (3.10 million acres) with 99.45 % occurring in counties on or west of the 100th meridian. To deal with these anomalies, it is assumed that zero acreage is available for terrestrial feedstock production from counties where irrigated pastureland exceeds combined permanent pasture and cropland pasture acreage. Further, the Census of Agriculture includes only private lands. Thus, federal lands are excluded from analysis in POLYSYS.

The amount of pastureland in the 2007 Census of Agriculture occurring in counties that lie east of the 100th meridian was 59 million hectares (146 million acres). Using the methodology described above to exclude (1) irrigated pastureland; (2) cropland pasture; and (3) counties where the occurrence of irrigated pastureland exceeds combined reported permanent pasture and cropland pasture, this acreage is discounted to 53.41 million hectares (132 million acres). In comparison with the national amount of reported pastureland and cropland pasture in farms, our methodology excludes roughly 68 % of the total US land classified in the USDA Census as "permanent pasture" or "cropland pasture" in farms. Further, as pasture is claimed by dedicated energy crop production livestock forage from grazing can be displaced. To maintain forage and livestock production, POLYSYS includes an assumption requiring that 1 ha of pastureland be converted to management-intensive grazing for every hectare of pasture converted to dedicated biomass production. As a result, the costs of establishment on pastureland are higher than for cropland due to the costs to intensify grazing (through fencing, fertilizer applications, rotation of cattle, etc.). These four conditions (non-wooded, nonirrigated, non-arid, and forage intensification) lower the total pastureland amount feasibly available for conversion to 26.5 million hectares (65.5 million acres) or about 16 % of the total US pastureland and cropland used as pasture. The biophysical and economic constraints in POLYSYS act together to reduce projected competition of pastureland among terrestrial and livestock systems as intended for the purpose of assessment within the Billion-Ton Update, as well as algal feedstocks, as the focus of this study.

Biomass crops are allocated to agricultural land based on relative profitability to conventional major crops. Residues are collected when there is positive profit. The model can be used to estimate feedstock availability in two ways. Either a specific biomass price or a specific biofuel demand level can be exogenously introduced to the model. If biomass price is exogenously introduced, all feedstocks in all regions are offered the same price per dry Mg, land is allocated to the most profitable regional feedstocks, and the model determines the national supply level associated with the exogenous price. If an exogenous demand for biofuels is introduced, then the model iterates by incrementally increasing the biomass price until the national demand level is reached. By introducing an exogenous demand level, the model determines the equilibrium feedstock price that is necessary to meet this demand. More information about POLYSYS assumptions and operations are available from De la Torre Ugarte and Ray (2000). Further documentation regarding the application of POLYSYS to biomass crops is available in the appendix of English et al. (2010).

2.3.2 *Woody Feedstock Modeling Assumptions in the Northeast*

Woody biomass resources simulated in POLYSYS include (1) SRWCs on agricultural land and (2) forest resources that may be derived from forestland. Following is a description of two broad feedstock categories and their assumptions specific to the 12 Northeast states.

2.3.2.1 Short-Rotation Woody Crops

SRWCs are simulated with one of two general management strategies: (1) a single-stem, non-coppicing, system managed on an 8-year rotation, designed to simulate poplar, pine, or eucalyptus production typical of systems in the US South and Northwest, and (2) a multi-stem, coppicing system, managed on a 4-year cutting cycle with 20 years between replanting, simulating shrub willow (*Salix* spp.) in the Northeast USA. Focusing on the Northeast, shrub willow is assumed to yield 6 dry tons per acre per year. Establishment costs of willow are assumed to be $1,120 per acre, second-year maintenance costs are assumed to be $30 per acre, and subsequent maintenance costs are assumed to be $100 per acre per year. Resulting discounted average costs of production are $38–$45 per dry ton. Costs assume a discount rate of 6.5 % and include all variable costs exclusive of land rent.

The BT2 evaluated two scenarios—baseline and high yield. Because dedicated feedstocks like willow are modeled to compete with conventional crops, the scenarios have assumptions for not just dedicated feedstock yields but also conventional crops. The baseline scenario assumes a continuation of the US Department of Agriculture forecasts for the major food and forage crops. Energy crop yields, including willow, assume an annual increase of 1 %. The 1 % change in annual yield in the baseline reflects learning or experience in planting energy crops and limited gains that can be had through breeding and selection of better varieties.

The high-yield scenario is more closely aligned to the assumptions in the 2005 BTS. In this scenario, higher corn yields and a much larger fraction of crop acres in reduced and no-till cultivation are assumed. The energy crop productivity increases, again including willow, are modeled at three levels—2, 3, and 4 % annually. These gains are due not only to experience in planting energy crops but also to more aggressive implementation of breeding and selection programs. High-yield assumptions are based on a series of workshops sponsored by the DOE to obtain information concerning the development of industry-based, high-yield alternatives to the baseline assumptions used to develop the update (U.S. Department of Energy 2009).

In sum, the woody crop modeled in the BT2 for the Northeast USA is willow, with baseline assumptions including 1 % annual yield improvements and three high-yield scenarios including annual yield improvement of 2, 3, and 4 % annually. More details about these modeling assumptions are available in DOE (2011).

2.3.2.2 Forest Resources

While POLYSYS simulates dedicated feedstocks (woody and herbaceous) and agricultural residues that might be procured from agricultural lands, the economic availability of woody resources is also evaluated in the BT2. Forest biomass and wood waste resources included in the report, and shown here for the Northeast USA, include the following (adapted from DOE (2011)):

Forest Residues (Logging Residues and Thinnings) from Integrated Forest Operations from Timberland[1]

Forest residues are a primary resource that consists of a combination of estimates from two sources: (1) removal of a portion of the logging residue that is currently

[1] Forestland is defined as land at least 120 ft wide and 1 ac in size, with at least 10 % cover (or equivalent stocking) by live trees of any size, including land that formerly had such tree cover and that will be naturally or artificially regenerated (Smith et al. 2009). Forestland is further defined as timberland and other forestland. Timberland is defined as forestland that is producing, or is capable of producing, in excess of 20 cubic feet per acre per year of industrial wood and not withdrawn from timber utilization by statute or administrative regulation. Other forestland is defined as forestland other than timberland and productive reserved forestland. It includes available forestland, which is incapable of annually producing 20 cubic feet per acre per year. Reserved forestland is administratively removed from production.

generated during the harvesting of timberlands for conventional forest products and (2) removal of excess biomass from fuel treatment (reducing biomass helps forests increase fire resistance) and thinning operations designed to reduce risks and losses from catastrophic fires and improve forest health. This latter component consists of removing merchantable whole trees and excess small trees to the roadside. The tops and branches of merchantable trees, cull trees, cull-tree components, and excess small trees can be used for bioenergy applications. The merchantable tree components can be used for conventional forest products. Both of these resources were considered separately in the BTS, but in this update, estimates are made assuming that there will be a transition in conventional harvesting operations from leaving logging residues behind to removing them as part of conventional harvesting. It is projected that access to biomass will come from integrated harvesting operations that provide sawlogs and pulpwood to meet the existing market demand as well as provide biomass for energy.

Other Removal Residue[2]

Other removal residue is unused wood that is cut during the conversion of timberland to non-forest uses and unused wood cut in silvicultural operations, such as precommercial thinnings.

Thinnings from Other Forestland

Thinnings from other forestland (nontimberland) are conducted to improve forest health by removing excess biomass on low-productivity land.

Unused Primary and Secondary Mill Processing Residues

The processing of sawlogs, pulpwood, and veneer logs into conventional forest products generates significant quantities of bark, mill residues (coarse and fine wood), and pulping liquors. With the exception of small quantities of mill residues, these secondary forest product industry residues are currently used in the manufacture of forest products or for heat and power production, and valuable chemicals are recovered from pulping liquors. In addition to pulping liquors, fuelwood—defined as wood harvested directly from forests and used primarily in the residential and commercial sectors for space heating and by some electric utilities for power

[2] Unutilized wood volume from cut, or otherwise killed, growing stock from cultural operations, such as precommercial thinnings or from timberland clearing. Does not include volume removed from inventory through reclassification of timber land to productive reserved forest land (Smith et al. 2009).

generation—is also not considered beyond the estimates provided in Chap. 1. Some quantity of these currently used wood wastes could shift to bioenergy applications at the right price. However, estimating how many of these resources could move into bioenergy production is difficult and speculative, as many of these wood wastes are not only used but are also confined or dedicated to a specific process.

Urban Wood Wastes

Urban wood waste is largely destined for landfills. The urban wood waste resource includes a wide variety of woody materials, ranging from discarded furniture, landscaping wood wastes, and wood used in the construction, remodeling, and demolition of buildings.

Conventionally Sourced Wood

Conventionally sourced wood is defined as separate operations to provide pulpwood-sized roundwood for bioenergy applications. Conventional wood was not included in the 2005 *BTS*. Excluded from the forest potential is wood grown under short rotations and dedicated to bioenergy production.

See DOE (2011) Chap. 5 for detailed modeling assumptions of these resources.

2.4 Results

Detailed POLYSYS outputs used in the Billion-Ton Update (U.S. Department of Energy 2011) for all resources, counties, years, and scenarios are available at www.bioenergykdf.net. Here results are summarized with focus on the Northeast USA. All supplies are reported as potentially available at the farmgate or the forest landing.

The majority of woody feedstocks in the Northeast are expected to come from integrated forest operations and SRWCs totaling 5.5 and 7.5 million Mg, respectively, assuming a forest landing price of \$66 dry Mg^{-1} in 2022 (Fig. 2.2 and Table 2.11). Forest resources are projected to come primarily from states with established forest industries, including Pennsylvania, Maine, and West Virginia (Table 2.2), while SRWCs are likely to come largely from states with a significant agricultural base, such as Pennsylvania and New York (Table 2.1). The projected availability of feedstock increases with farmgate (or forest landing) price. The proportion of states' contributions to supply remains fairly constant with increasing farmgate price, with West Virginia, Pennsylvania, New York, and Maine collectively providing the majority of feedstocks (Fig. 2.3). At a farmgate price lower than \$50 dry Mg^{-1}, feedstock supply is largely comprised of forest residues and thinnings from integrated operations, while prices between \$50–\$60 dry Mg^{-1} incentivize the

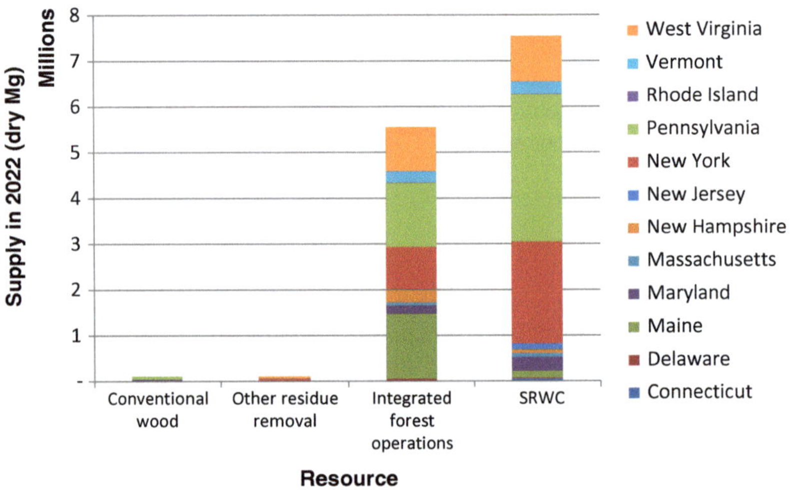

Fig. 2.2 Woody biomass feedstocks projected in 2022, assuming a farmgate price of \$66 dry Mg^{-1} in 12 Northeast states. Data accessed from the Bioenergy knowledge discovery framework, https://bioenergykdf.net. 2/15/2013

Table 2.1 SRWC supply (thousand dry metric tons) by price and year

Year	2012				2022				2030			
Farmgate price (\$ dry Mg^{-1})	\$22	\$44	\$66	\$88	\$22	\$44	\$66	\$88	\$22	\$44	\$66	\$88
State												
Connecticut	–	–	–	–	–	–	62	54	–	–	107	90
Delaware	–	–	–	–	–	–	19	17	–	–	32	171
Maine	–	–	–	–	–	–	148	135	–	–	249	224
Maryland	–	–	–	–	–	–	308	313	–	–	534	493
Massachusetts	–	–	–	–	–	–	84	74	–	–	146	125
New Hampshire	–	–	–	–	–	–	76	67	–	–	128	114
New Jersey	–	–	–	–	–	–	134	142	–	–	229	217
New York	–	–	–	–	–	–	2,221	2,040	–	9	3,355	2,982
Pennsylvania	–	–	–	–	–	–	3,224	3,866	–	–	4,859	5,002
Rhode Island	–	–	–	–	–	–	8	6	–	–	16	13
Vermont	–	–	–	–	–	–	271	255	–	–	462	426
West Virginia	–	–	–	–	–	–	989	1,030	–	70	1,447	1,416
Northeast total	–	–	–	–	–	–	7,542	7,999	–	79	11,565	11,272
Remaining USA	–	–	–	–	–	–	68,822	118,090	–	–	102,468	155,547
Contiguous USA	–	–	–	–	–	–	76,363	126,089	–	79	114,034	166,819

Supplies are cumulative. Data accessed from the Bioenergy knowledge discovery framework, https://bioenergykdf.net. 2/15/2013

production of SRWCs, and conventional roundwood is potentially available at prices greater than \$70 dry Mg^{-1} (Fig. 2.4). State-level, resource-specific supply projections by price and year are shown in Tables 2.1, 2.2, 2.3, 2.4, 2.5, 2.6, 2.7, 2.8, 2.9, 2.10, and 2.11.

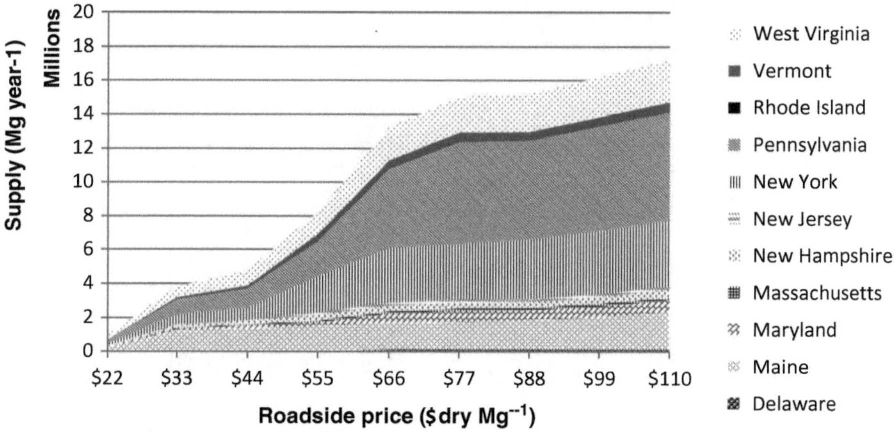

Fig. 2.3 Woody biomass feedstock supplies projected in 2022 by state, in 12 Northeast states. Feedstocks include residues and thinnings from integrated operations on timberland, other forestland thinnings, conventional wood, and SRWCs. Data accessed from the Bioenergy knowledge discovery framework, https://bioenergykdf.net. 2/15/2013

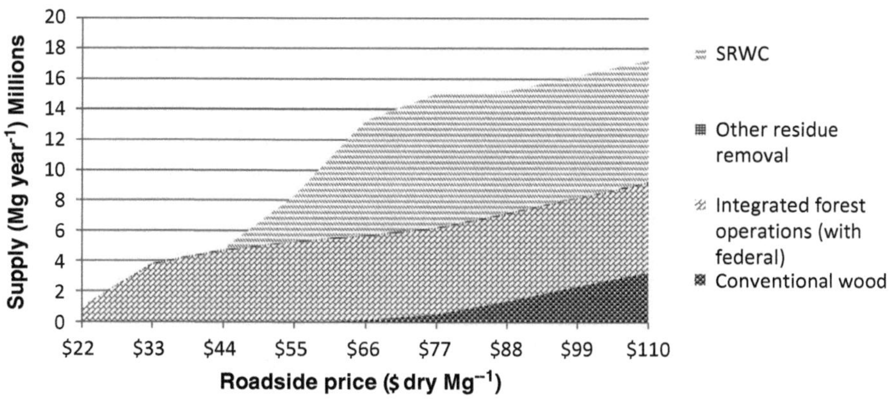

Fig. 2.4 Woody biomass feedstocks projected in 2022 in 12 Northeast states by feedstock type. Data accessed from the Bioenergy knowledge discovery framework, https://bioenergykdf.net. 2/15/2013

The 12 Northeast states comprise about 7 % of the area of the contiguous US states and represent a comparable amount of the total feedstock supply projected by DOE (2011). Depending on assumed feedstock price and year, national production of biomass from agricultural lands (e.g., agricultural residues, and dedicated herbaceous crops and SRWCs) is estimated to range from 54 million dry Mg (assuming

Table 2.2 Forest residue supply (thousand dry metric tons) by price and year, including federal lands

Year	2012				2022				2030			
Farmgate price ($ dry Mg^{-1})	$22	$44	$66	$88	$22	$44	$66	$88	$22	$44	$66	$88
State												
Connecticut	1	19	19	19	1	19	19	19	1	19	19	19
Delaware	5	34	39	43	5	33	39	42	5	33	38	42
Maine	273	1,345	1,447	1,543	271	1,322	1,424	1,520	271	1,319	1,421	1,517
Maryland	30	187	187	187	30	185	185	185	30	185	185	185
Massachusetts	7	60	60	60	6	58	58	58	7	58	58	58
New Hampshire	32	212	280	280	32	210	278	278	33	211	279	279
New Jersey	–	9	9	9	–	9	9	9	–	9	9	9
New York	108	700	934	1,046	108	694	928	1,040	108	694	928	1,040
Pennsylvania	241	1,112	1,394	1,394	244	1,119	1,400	1,400	246	1,125	1,407	1,407
Rhode Island	0	5	5	5	0	5	5	5	0	5	5	5
Vermont	34	184	246	263	33	182	244	261	33	182	244	261
West Virginia	252	848	960	960	256	855	968	968	258	861	974	974
Northeast total	982	4,712	5,579	5,807	986	4,691	5,556	5,784	992	4,699	5,566	5,794
Remaining USA	7,663	27,843	30,539	32,053	7,849	28,591	31,289	32,804	7,897	28,880	31,575	33,091
Contiguous USA	8,646	32,555	36,118	37,860	8,834	33,282	36,845	38,589	8,889	33,579	37,141	38,886

Supplies are cumulative. Forest residues include a fraction of both logging residues and thinnings from integrated forest operations from timberland. Data accessed from the Bioenergy knowledge discovery framework, https://bioenergykdf.net. 2/15/2013

Table 2.3 Other removal residues (thousand dry metric tons) by price and year, including federal lands

Year	2012				2022				2030			
Farmgate price ($ dry Mg^{-1})	$22	$44	$66	$88	$22	$44	$66	$88	$22	$44	$66	$88
State												
Connecticut	–	0	0	0	–	0	0	0	–	0	0	0
Delaware	–	–	–	–	–	–	–	–	–	–	–	–
Maine	–	0	0	0	–	0	0	0	–	0	0	0
Maryland	0	1	1	1	0	1	1	1	0	1	1	1
Massachusetts	–	–	–	–	–	–	–	–	–	–	–	–
New Hampshire	–	0	0	0	–	0	0	0	–	0	0	0
New Jersey	–	–	–	–	–	–	–	–	–	–	–	–
New York	17	68	68	68	17	68	68	68	17	69	69	69
Pennsylvania	–	0	0	0	–	0	0	0	–	0	0	0
Rhode Island	–	–	–	–	–	–	–	–	–	–	–	–
Vermont	–	–	–	–	–	–	–	–	–	–	–	–
West Virginia	0	1	1	1	0	1	1	1	0	1	1	1
Northeast total	17	71	71	71	17	71	71	71	18	72	72	72
Remaining USA	3,932	11,219	11,219	11,219	3,950	11,273	11,273	11,273	3,964	11,317	11,317	11,317
Contiguous USA	3,950	11,290	11,290	11,290	3,967	11,344	11,344	11,344	3,982	11,389	11,389	11,389

Supplies are cumulative. Data accessed from the Bioenergy knowledge discovery framework, https://bioenergykdf.net. 2/15/2013

Table 2.4 Thinnings from other forestland (nontimberland) (thousand dry metric tons) by price and year, including federal lands

Year	2012				2022				2030			
Farmgate price ($ dry Mg^{-1})	$22	$44	$66	$88	$22	$44	$66	$88	$22	$44	$66	$88
State												
Connecticut	–	–	–	–	–	–	–	–	–	–	–	–
Delaware	–	–	–	–	–	–	–	–	–	–	–	–
Maine	–	–	–	–	–	–	–	–	–	–	–	–
Maryland	–	–	–	–	–	–	–	–	–	–	–	–
Massachusetts	–	–	–	–	–	–	–	–	–	–	–	–
New Hampshire	–	–	–	–	–	–	–	–	–	–	–	–
New Jersey	–	–	–	–	–	–	–	–	–	–	–	–
New York	–	–	–	–	–	–	–	–	–	–	–	–
Pennsylvania	–	–	–	–	–	–	–	–	–	–	–	–
Rhode Island	–	–	–	–	–	–	–	–	–	–	–	–
Vermont	–	–	–	–	–	–	–	–	–	–	–	–
West Virginia	–	–	–	–	–	–	–	–	–	–	–	–
Northeast total	–	–	–	–	–	–	–	–	–	–	–	–
Remainder of lower 48	–	–	5,800	11,608	–	–	5,800	11,608	–	–	5,800	11,608
Contiguous USA	–	–	5,800	11,608	–	–	5,800	11,608	–	–	5,800	11,608

Supplies are cumulative. Data accessed from the Bioenergy knowledge discovery framework, https://bioenergykdf.net. 2/15/2013

$44.00 dry Mg^{-1} in 2012) to 603 million dry Mg (assuming $66.00 dry Mg^{-1} in 2030).[3] The 12 Northeast states are estimated to produce from 1 to 4 % of this nationwide biomass supply from agricultural lands. National production of feedstocks from forestlands (e.g., logging residues, thinnings, and processing wastes) ranges from about 72 million dry Mg (assuming a forest landing price of $44.00 dry Mg^{-1} in 2012) to 93 million dry Mg (assuming a forest landing price of $66.00 dry Mg^{-1} in 2030). The 12 Northeast states produce about 4–12 % of this national supply under a wide range ($11.00–$88.00 dry Mg^{-1}) of forest landing prices. Thus, the strong forest industry in the Northeast is represented in national supplies. Agriculture also plays a role, though not as significant as in the Midwest and plains states.

Forest residue supplies reported here are more conservative than results reported by some state-level projections. For example, Walker et al. (2010) report that between 68,000 and 401,000 dry Mg $year^{-1}$ of forest biomass is potentially available in Massachusetts, depending on price scenario. This is an order of magnitude higher than the 7–60 dry Mg $year^{-1}$ of forest residues reported in Table 2.2. However, with

[3] All prices are at the farmgate. These projections are derived from the base-case scenario.

Table 2.5 Unused primary mill residues (thousand dry metric tons) by price and year

Year	2012				2022				2030			
Farmgate price ($ dry Mg^{-1})	$22	$44	$66	$88	$22	$44	$66	$88	$22	$44	$66	$88
State												
Connecticut	4	4	4	4	4	4	4	4	4	4	4	4
Delaware	–	–	–	–	–	–	–	–	–	–	–	–
Maine	38	38	38	38	38	38	38	38	39	39	39	39
Maryland	16	16	16	16	15	15	15	15	16	16	16	16
Massachusetts	–	–	–	–	–	–	–	–	–	–	–	–
New Hampshire	6	6	6	6	6	6	6	6	6	6	6	6
New Jersey	1	1	1	1	1	1	1	1	1	1	1	1
New York	25	25	25	25	25	25	25	25	25	25	25	25
Pennsylvania	155	155	155	155	155	155	155	155	156	156	156	156
Rhode Island	0	0	0	0	0	0	0	0	0	0	0	0
Vermont	–	–	–	–	–	–	–	–	–	–	–	–
West Virginia	108	108	108	108	107	107	107	107	108	108	108	108
Northeast total	353	353	353	353	353	353	353	353	355	355	355	355
Remaining USA	832	832	832	832	888	888	888	888	925	925	925	925
Contiguous USA	1,186	1,186	1,186	1,186	1,241	1,241	1,241	1,241	1,281	1,281	1,281	1,281

Supplies are cumulative. Data accessed from the Bioenergy knowledge discovery framework, https://bioenergykdf.net. 2/15/2013

Table 2.6 Unused secondary mill residues (thousand dry metric tons) by price and year

Year	2012–2030			
Farmgate price ($ dry Mg^{-1})	$22	$44	$66	$88
State				
Connecticut	29	29	29	29
Delaware	6	6	6	6
Maine	62	62	62	62
Maryland	44	44	44	44
Massachusetts	45	45	45	45
New Hampshire	63	63	63	63
New Jersey	21	21	21	21
New York	8	8	8	8
Pennsylvania	219	219	219	219
Rhode Island	7	7	7	7
Vermont	165	165	165	165
West Virginia	165	165	165	165
Northeast total	834	834	834	834
Remaining USA	4,700	4,700	4,700	4,700
Contiguous USA	5,534	5,534	5,534	5,534

Supplies are cumulative. Data accessed from the Bioenergy knowledge discovery framework, https://bioenergykdf.net. 2/15/2013

the addition of other woody sources, primarily SRWCs and urban wood waste, combined woody biomass supplies reported here range from 277,000 to 839,000 dry Mg (Table 2.10), more than twice the amount reported by Walker et al. (2010). For the state of New York, Wojnar (2012) reported between 3.9 and 7.6 million dry Mg $year^{-1}$ of forest biomass as potentially available annually, more than the 0.1–1.0 million Mg of forest residues reported in Table 2.2. This may be attributable to the economic and environmental constraints assumed for forest residues in the US Department of Energy report (2011). Quantities estimated by Wojnar (2012) are within the 0.8–6.1 million dry Mg $year^{-1}$ of total woody biomass supplies reported in Table 2.10. Thus, estimates of potential supplies from forest lands reported here may be conservative, with dedicated feedstocks like SRWCs showing great potential at prices greater than about $50 dry Mg^{-1} at the farmgate.

2.5 Summary and Conclusions

Biomass feedstocks in the Northeast USA are projected to be more available both with higher feedstock prices and over time, particularly as SRWC plantations come into production. Under baseline projections as defined by DOE (2011),

Table 2.7 Urban wood waste fraction of construction and demolition debris (thousand dry metric tons) by price and year

Year	2012				2022				2030			
Farmgate price ($ dry Mg^{-1})	$22	$44	$66	$88	$22	$44	$66	$88	$22	$44	$66	$88
State												
Connecticut	45	145	223	223	43	139	214	214	42	137	211	211
Delaware	11	37	57	57	12	37	57	57	12	38	58	58
Maine	17	57	87	87	18	58	88	88	18	59	91	91
Maryland	76	248	381	381	81	262	403	403	84	274	421	421
Massachusetts	83	269	414	414	81	263	404	404	81	262	404	404
New Hampshire	18	59	90	90	19	62	95	95	20	64	99	99
New Jersey	112	364	560	560	109	354	544	544	108	351	540	540
New York	242	787	1,211	1,211	229	744	1,145	1,145	224	727	1,118	1,118
Pennsylvania	158	514	791	791	152	495	761	761	150	488	750	750
Rhode Island	14	45	69	69	13	43	66	66	13	43	67	67
Vermont	8	27	42	42	9	29	44	44	9	29	45	45
West Virginia	23	76	116	116	22	73	113	113	22	73	112	112
Northeast total	807	2,628	4,041	4,041	787	2,559	3,935	3,935	783	2,546	3,915	3,915
Remaining USA	3,205	10,472	16,085	16,085	3,445	11,251	17,283	17,283	3,662	11,950	18,358	18,358
Contiguous USA	4,012	13,100	20,126	20,126	4,233	13,810	21,217	21,217	4,445	14,496	22,273	22,273

Supplies are cumulative. Data accessed from the Bioenergy knowledge discovery framework, https://bioenergykdf.net. 2/15/2013

Table 2.8 Urban wood waste fraction of municipal solid waste (thousand dry metric tons) by price and year

Year	2012				2022				2030			
Farmgate price ($ dry Mg^{-1})	$22	$44	$66	$88	$22	$44	$66	$88	$22	$44	$66	$88
State												
Connecticut	77	92	102	102	74	88	98	98	73	87	97	97
Delaware	20	23	26	26	20	24	26	26	20	24	27	27
Maine	30	36	40	40	31	37	41	41	31	37	41	41
Maryland	132	158	175	175	139	167	185	185	146	175	194	194
Massachusetts	143	171	190	190	140	167	186	186	140	167	186	186
New Hampshire	31	37	41	41	33	39	44	44	34	41	45	45
New Jersey	194	232	258	258	188	226	250	250	187	224	248	248
New York	419	502	558	558	396	474	527	527	386	463	514	514
Pennsylvania	274	327	363	363	263	315	350	350	260	311	345	345
Rhode Island	24	28	32	32	23	28	31	31	23	28	31	31
Vermont	15	17	19	19	15	18	20	20	16	19	21	21
West Virginia	40	48	53	53	40	48	52	52	39	47	51	51
Northeast total	1,397	1,674	1,858	1,858	1,361	1,631	1,809	1,809	1,354	1,622	1,800	1,800
Remaining US	5,576	6,682	7,393	7,393	5,993	7,180	7,947	7,947	6,362	7,625	8,443	8,443
Contiguous USA	6,974	8,355	9,251	9,251	7,354	8,810	9,757	9,757	7,716	9,247	10,243	10,243

Supplies are cumulative. Data accessed from the Bioenergy knowledge discovery framework, https://bioenergykdf.net. 2/15/2013

Table 2.9 Conventional wood available for bioenergy (thousand dry metric tons) by price and year

Year	2012				2022				2030			
Farmgate price ($ dry Mg^{-1})	$22	$44	$66	$88	$22	$44	$66	$88	$22	$44	$66	$88
State												
Connecticut	–	–	–	0	–	–	–	1	–	–	–	1
Delaware	–	–	–	14	–	–	–	17	–	–	–	18
Maine	–	–	–	63	–	–	–	75	–	–	–	81
Maryland	–	–	36	66	–	–	43	78	–	–	47	84
Massachusetts	–	–	–	6	–	–	–	8	–	–	–	8
New Hampshire	–	–	–	–	–	–	–	–	–	–	–	–
New Jersey	–	–	–	–	–	–	–	–	–	–	–	–
New York	–	–	–	301	–	–	–	357	–	–	–	387
Pennsylvania	–	–	52	503	–	–	61	596	–	–	66	647
Rhode Island	–	–	–	1	–	–	–	1	–	–	–	1
Vermont	–	–	–	–	–	–	–	–	–	–	–	–
West Virginia	–	–	–	181	–	–	–	214	–	–	–	232
Northeast total	–	–	88	1,136	–	–	104	1,345	–	–	113	1,459
Remaining USA	–	–	1,261	15,895	–	–	1,494	18,826	–	–	1,621	20,429
Contiguous USA	–	–	1,349	17,031	–	–	1,598	20,171	–	–	1,734	21,888

Supplies are cumulative. Data accessed from the Bioenergy knowledge discovery framework, https://bioenergykdf.net. 2/15/2013

woody biomass production in the 12 Northeast US states ranges from 4.3 million dry Mg at a price of $22 dry Mg^{-1} in 2012 to 25.5 million dry Mg at a price of $88 dry Mg^{-1} in 2030 (Fig. 2.5). Initial woody biomass supplies in the Northeast are expected to be comprised almost entirely of nearly equal proportions of urban wood waste and integrated forest operations, with potential for significant supplies from SRWCs starting around 2022. Supplies from conventional wood and other residue removals are comparatively small (Fig. 2.6 and Table 2.10). Actual future supplies will be driven by factors such as regional and global economies, market conditions, bioenergy conversion technologies and associated feedstock specifications, state and national energy policies, and climate and extreme weather events.

Table 2.10 Combined woody biomass supplies (SRWC, forest residues, other removal residues, thinnings from other forestland, unused primary and secondary mill residues, urban wood waste from C&D and MSW, and conventional wood) available for bioenergy (thousand dry metric tons) by price and year

Year	2012				2022				2030			
Farmgate price ($ dry Mg^{-1})	$22	$44	$66	$88	$22	$44	$66	$88	$22	$44	$66	$88
State												
Connecticut	155	288	376	377	150	279	425	417	149	276	467	450
Delaware	41	100	128	146	42	100	147	166	42	101	162	322
Maine	422	1,540	1,676	1,835	420	1,518	1,802	1,960	421	1,517	1,904	2,056
Maryland	298	653	840	870	310	674	1,184	1,224	320	694	1,442	1,438
Massachusetts	277	544	708	714	271	533	776	775	271	532	839	826
New Hampshire	151	377	481	481	154	381	563	554	156	385	620	606
New Jersey	328	627	848	848	320	611	960	967	317	606	1,048	1,036
New York	818	2,090	2,803	3,216	782	2,013	4,920	5,209	768	1,994	6,017	6,142
Pennsylvania	1,046	2,328	2,974	3,425	1,033	2,304	6,170	7,347	1,031	2,300	7,802	8,526
Rhode Island	45	85	113	113	44	83	118	116	44	83	126	123
Vermont	222	393	473	489	222	393	744	745	223	395	937	918
West Virginia	588	1,245	1,403	1,584	591	1,249	2,395	2,650	593	1,325	2,859	3,060
Northeast total	4,391	10,271	12,823	14,099	4,338	10,138	20,203	22,130	4,335	10,208	24,221	25,502
Remaining USA	25,909	61,748	77,829	99,786	26,825	63,883	149,496	223,419	27,510	65,396	185,208	264,419
Contiguous USA	30,300	72,019	90,653	113,884	31,163	74,021	169,700	245,549	31,845	75,604	209,429	289,921

Supplies are cumulative. Data accessed from the Bioenergy knowledge discovery framework, https://bioenergykdf.net. 2/15/2013

Table 2.11 Cumulative woody biomass supplies (thousand dry metric tons) by price and year for the 12 Northeast states

Year	2012				2022				2030			
Farmgate price ($ dry Mg^{-1})	$22	$44	$66	$88	$22	$44	$66	$88	$22	$44	$66	$88
Feedstock												
SRWC	–	–	–	–	–	–	7,542	7,999	–	79	11,565	11,272
Integrated forest operations (with federal)	982	4,712	5,579	5,807	986	4,691	5,556	5,784	992	4,699	5,566	5,794
Other residue removal	17	71	71	71	17	71	71	71	18	72	72	72
Mill residue, unused primary, state	353	353	353	353	353	353	353	353	355	355	355	355
Mill residue, unused secondary	834	834	834	834	834	834	834	834	834	834	834	834
Urban woodwaste, construction and demolition, state	807	2,628	4,041	4,041	787	2,559	3,935	3,935	783	2,546	3,915	3,915
Urban wood waste, municipal solid waste, state	1,397	1,674	1,858	1,858	1,361	1,631	1,809	1,809	1,354	1,622	1,800	1,800
Conventional wood	–	–	88	1,136	–	–	104	1,345	–	–	113	1,459
Total for the Northeast	4,391	10,271	12,823	14,099	4,338	10,138	20,203	22,130	4,335	10,208	24,221	25,502
Total for the remaining lower 48	25,909	61,748	77,829	99,786	26,825	63,883	149,496	223,419	27,510	65,396	185,208	264,419
Total for the contiguous USA	30,300	72,019	90,653	113,884	31,163	74,021	169,700	245,549	31,845	75,604	209,429	289,921

Supplies are cumulative. Data accessed from the Bioenergy knowledge discovery framework, https://bioenergykdf.net. 2/15/2013

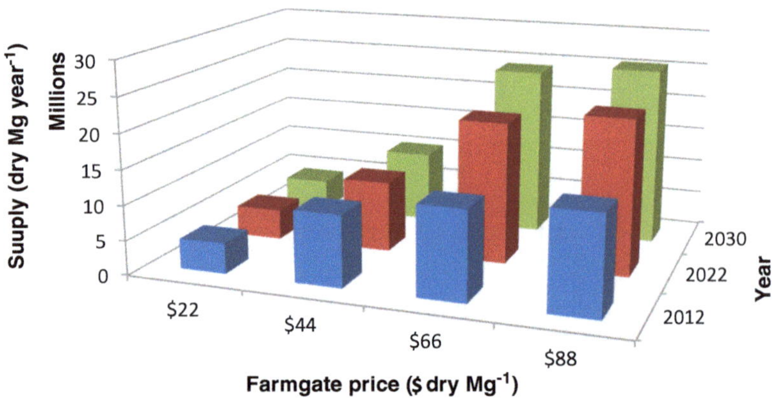

Fig. 2.5 Economic availability of woody biomass feedstocks for the 12 Northeast US states. Data accessed from the Bioenergy knowledge discovery framework, https://bioenergykdf.net. 2/15/2013

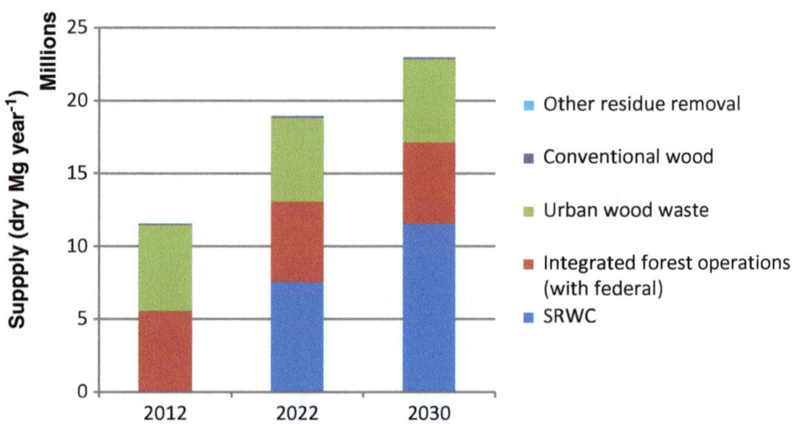

Fig. 2.6 Supplies of woody biomass feedstocks for the 12 Northeast US states assuming a farmgate price of \$66 dry Mg^{-1}. Data accessed from the Bioenergy knowledge discovery framework, https://bioenergykdf.net. 2/15/2013

References

Aliston JM, Norton G, Pardey PG (1995) Science under scarcity: principles and practice for agricultural research evaluation and priority setting. Cornell University Press, Ithinca, NY

De la Torre UD, Ray DE (2000) Biomass and bioenergy applications of the POLYSYS modeling framework. Biomass Bioenerg 18(4):291–308, <Go to ISI>://000086604500005

English B, De la Torre Ugarte D, Hellwinckel C, Jensen K, Menard RJ, West TO, Clark C (2010) Implications of energy and carbon policies for the agriculture and forestry sectors. Department of Agricultural and Resource Economics, Institute of Agriculture, The University of Tennessee, Knoxville, TN

Khanna M, Chen X, Huang H, Önal H (2011) Supply of cellulosic biofuel feedstocks and regional production pattern. Am J Agric Econ 93(2):7

Langholtz M, Graham RC, Eaton L, Perlack R, Hellwinckel C, De la Torre Ugarte D (2012) Price projections of feedstocks for biofuels and biopower in the U.S. Energy Policy, 41, 9. http://authors.elsevier.com/offprints/JEPO5395/7d6eee227fc4c5e6a2e78127a7590e62

Lin W, Westcott PC, Skinner R, Sanford S, De la Torre UD (2000) Supply response under the 1996 Farm Act and implications for the U.S. field crops sector. Market and Trade Economics Division, Economic Research Service, United States Department of Agriculture, Washington, DC

Perlack R, Wright L, Turhollow AF, Graham RL, Stokes B, Erback D (2005) Biomass as feedstock for a bioenergy and bioproducts industry: the technical feasibility of a billion-ton annual supply. Oak Ridge National Laboratory, Oak Ridge, TN, pp 1–78, http://feedstockreview.ornl.gov/pdf/billion_ton_vision.pdf#search='Biomass as feedstock for a bioenergy and bioproducts industry: the technical feasibility of a billionton annual supply'

Piggot RR (1992) Some old truths revisited. Aust J Agric Econ 26(2):117–140

Ray D, De la Torre UD, Dicks MR, Tiller K (1998) The POLYSYS modeling framework: a documentation. Agricultural Policy Analysis Center, University of Tennessee, Knoxville, TN, www.agpolicy.org/polysys.html

Smith W, Miles P, Perry CH, Pugh S (2009) Forest resources of the United States, 2007, U.S. Department of Agriculture Forest Service. Washington, DC

U.S. Department of Energy (2009). High-yield scenario workshop series (No. INL/EXT-10-20074). p 125

U.S. Department of Energy (2011) U.S. billion-ton update: biomass supply for a bioenergy and bioproducts industry. Oak Ridge National Laboratory, Oak Ridge, TN, pp 1–227, http://feedstockreview.ornl.gov/pdf/billion_ton_vision.pdf#search='Biomass as feedstock for a bioenergy and bioproducts industry: the technical feasibility of a billionton annual supply'

USDOE (2012) Biomass multi-year program plan, November 2012. USDOE, Washington, DC, p 187

Walker T, Cardellichio P, Colnes A, Gunn J, Kittler B, Perschel B, Recchia C, Saah D (2010). Biomass sustainability and carbon policy study (No. NCI-2010-03). Manomet Center for Conservation Sciences, Boston, Massachusetts. http://www.mass.gov/eea/docs/doer/renewables/biomass/manomet-biomass-report-full-hirez.pdf

Wojnar Z (2012) Renewable fuels roadmap and sustainable biomass feedstock supply for New York, vol 2. NYSDERA, Albany, NY, p 24, http://www.nyserda.ny.gov/Publications/Research-and-Development-Technical-Reports/Biomass-Reports/Renewable-Fuels-Roadmap.aspx

Chapter 3
Carbon Cycle Implications of Forest Biomass Energy Production in the Northeastern United States

Charles D. Canham

3.1 Introduction

There has been enormous interest in the increased use of forest biomass as a component of the renewable energy portfolio of the northeastern United States. While often touted as an inherently "carbon-neutral" energy source, it has become clear that a wide range of factors need to be considered to evaluate the net carbon and climate impact of biomass energy production (Cook and Beyea 2000; Abbas et al. 2011; Schulze et al. 2012; Zanchi et al. 2012). Net growth in northeastern forests currently provides an important offset to fossil fuel emissions of greenhouse gases, but most studies predict that carbon sequestration rates in forests both regionally and nationwide will decline over the next 20–50 years (Hurtt et al. 2002; USDA Forest Service 2012). The decline in the magnitude of the annual net carbon sequestration in the US forests is believed to be a product of three main factors: a stabilizing or even reversal of the century-long trend of an increasing forestland base due to reforestation of abandoned farmland, as well as parcelization due to land development (Drummond and Loveland 2010), declining rates of net growth at the stand level because of forest maturation (Hurtt et al. 2002), and increasing rates of forest harvest (Radeloff et al. 2006). All of these raise important questions about the magnitude of the sustainable supply of biomass feedstocks from northeastern forests (Buchholz et al. 2011) as well as questions about the future of regional forest carbon stocks under new demand for biomass energy.

The assessment of net carbon emissions to the atmosphere from harvest and combustion of forest biomass feedstocks—termed here "net carbon impact"—is at heart an issue of ecosystem mass balance and carbon cycling. A net change in atmospheric carbon as a direct result of forest biomass energy production inevitably implies a corresponding set of net changes in one or more forest carbon pools.

C.D. Canham (✉)
Cary Institute of Ecosystem Studies, Box AB, Millbrook, NY 12545, USA
e-mail: canhamc@caryinstitute.org

M. Jacobson and D. Ciolkosz (eds.), *Wood-Based Energy in the Northern Forests*,
DOI 10.1007/978-1-4614-9478-2_3,

It is worth noting that net changes to forest carbon stocks are likely to have ecological consequences beyond impacting atmospheric conditions, since forest carbon (both living and dead) provides important ecosystem services quite apart from the debate about global warming (Nelson et al. 2009). Thus, while much of the current debate about forest biomass energy has focused on its net carbon impact, the most important broader questions would appear to be (1) whether we can manage the transition from northeastern forests as a net carbon sink, but with relatively little biomass energy production, to forests that provide a sustainable, renewable energy source without depleting regional forest carbon stocks, and (2) how much energy could we reasonably expect to obtain sustainably from northeastern forests.

This chapter outlines issues that determine the net carbon impact of utilization of the two basic bioenergy feedstocks sourced from northeastern forests: roundwood and logging residues. Following this is a simple analysis of factors that determine the supply of forest biomass at a landscape level and discussion of the challenges and policy implications of trying to engineer a "soft landing" for northeastern forests—i.e., the transition from net carbon sequestration in forests, but with little biomass energy production, to stable regional forest carbon stocks with higher sustainable biomass energy production.

3.2 A Simplified Northeastern Forest Carbon Cycle

A simplified carbon cycle for northeastern forests focuses on inputs and losses from two key stocks of organic forest carbon: live woody biomass and detrital biomass in standing dead trees, downed wood (coarse woody debris (CWD)), the surface layer of organic matter on the forest floor, and organic matter incorporated in the mineral soil (Fig. 3.1). Change in the stock of live woody biomass (at either the stand level or at a regional scale) is simply the balance of inputs via "growth" (net annual woody biomass increment) and annual losses via mortality. Mortality is typically partitioned into either "removals" in harvested trees—whether used for traditional

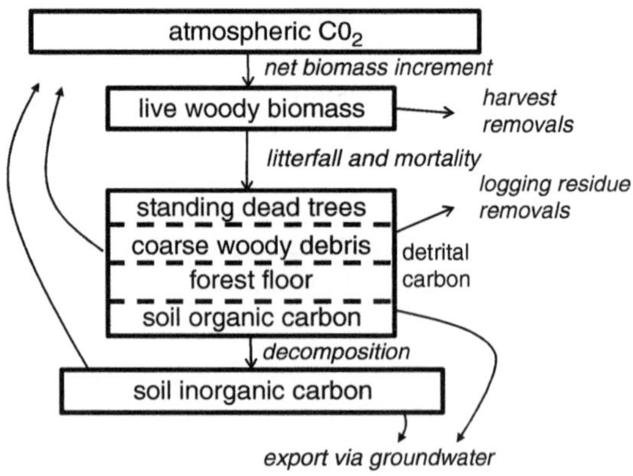

Fig. 3.1 A simplified forest carbon cycle, focusing on carbon in live woody biomass and detrital pools

forest products or as an energy feedstock—or all other forms of mortality combined. This simplified model ignores seasonal variation in non-woody biomass associated with leaf production and leaf fall. Live biomass in herbaceous and shrub species is also typically ignored in the analyses of carbon cycle implications of forest feedstock utilization. Many field studies include soil microbial biomass in estimates of forest floor and soil organic carbon, but those dynamics are ignored here as well. The detrital carbon pools differ in both size and rate of turnover. For example, organic carbon pools in the mineral soil can be quite large but typically have very long residence times.

It is useful to partition the inputs of detrital biomass into two subsets with very different time scales of inputs and losses: (1) a "background" input of leaf and fine branch litter to the forest floor, characterized by an annual cycle of relatively rapid turnover, and (2) an "episodic" input of CWD as a result of tree mortality or a disturbance that causes damage to tree crowns (i.e., wind or ice storms). This latter input is characterized by slower decomposition and much longer time transients. Analyses of the impacts of forest biomass energy projects are typically most concerned with these episodic inputs (e.g., Mika and Keeton 2012). Decomposition represents the major loss from detrital pools in northeastern forests, although combustion via fire can also be a factor. Although rarely explicitly accounted for, forests export a small quantity of the detrital pool as dissolved and particulate organic carbon in groundwater each year (typically 30–50 kg/ha/yr of dissolved organic carbon for forests in the northeastern United States, Canham et al. 2004).

3.3 Net Carbon Impact of Utilization of Roundwood Feedstocks: The Importance of Baselines

The argument that use of roundwood (or more generally, live woody biomass) as a feedstock can provide a renewable and "carbon-neutral" energy source is based on the premise that—in sustainably managed forests—carbon released to the atmosphere via feedstock combustion will be taken back out of the atmosphere as forests regrow. More detailed analyses, however, have led to widely disparate conclusions about the net carbon impact of the use of roundwood feedstocks, as a result of differences in assumptions about the appropriate "baseline" against which to assess carbon neutrality (Searchinger et al. 2009; Manomet Center 2010).

Perhaps the simplest formulation of a baseline has been termed the "reference point" approach (U.S. Environmental Protection Agency 2011). As long as regional forest carbon stocks (both live woody carbon and detrital carbon) are stable or increasing, forest feedstock utilization is considered to result in no net addition of CO_2 to the atmosphere. This approach requires an arbitrary decision about what the reference point should be when carbon stocks in a region are increasing. For instance, carbon stocks in northeastern forests steadily increased during the twentieth century (Smith et al. 2009). If the reference point had been fixed at forest carbon stock levels present at some arbitrary time during the past century, then presumably the carbon sequestered in northeastern forests since then could be consumed for use as a feedstock, and as long as total carbon stocks equilibrated at or above the level fixed in the past, the new feedstock use would be considered carbon neutral. But if a reference point of the year 2010 was used, and the region eventually equilibrated at forest carbon stocks below those in the year 2010, feedstock utilization would have resulted in a net addition of CO_2 to the atmosphere, *relative to the reference year.* This suggests that if a reference point baseline is used for calculation of net carbon impact then the baseline should be periodically revised upward if carbon stocks in a region increase.

Searchinger et al. (2009) highlights these issues and outlines what they consider to be a "critical climate accounting error" in existing IPCC standards for tracking net changes in carbon stocks. They argue that the appropriate baseline for assessment of the net carbon impact of a change in forest management is the expected time course of carbon storage that would have occurred in the absence of the new management. Given that most forest regions of the United States currently act as net carbon sinks, their approach would lead to the conclusion that any activity that reduces the rate of net carbon sequestration would, in effect, act as a net carbon source to the atmosphere, unless the feedstock utilization resulted in substitution for fossil fuel carbon emissions that more than outweighed the reduction in forest sink strength. This approach has been termed a "business as usual" (BAU) or "anticipated future" baseline. It begs the question of predicting exactly how the size of the carbon sink in the forestry sector will change in the future.

The BAU baseline establishes a very stringent standard for carbon neutrality. It effectively requires balancing the potential for further sequestration that would happen in the absence of changes to current forest management with the potential to offset fossil fuel carbon emissions through the use of biomass as a feedstock. As such, the calculation of net carbon impact from biomass energy under a BAU baseline requires specification of the technologies used for production of both the biomass energy and the fossil fuel energy. This then forms the basis for calculation of the tons of CO_2 emissions from fossil fuel consumption that would be offset by whatever tons of biomass feedstock were needed to produce an equivalent amount of useable energy (e.g., Manomet Center 2010). Because of the relatively low energy density of biomass feedstocks, and often low efficiencies of conversion to useable energy, the adoption of a BAU baseline emphasizes that in a region where forests are currently acting as a significant net sink for carbon, policies favoring sequestration (particularly as an offset to fossil fuel emissions) are likely to have a more positive effect on atmospheric CO_2 concentrations than policies favoring feedstock harvests and substitution for fossil fuel consumption.

From a practical perspective, there are enormous difficulties with implementing a BAU baseline in a carbon accounting framework. Not the least is the ability to accurately project the future time course of forest carbon stocks. A recent analysis of the accuracy of the decadal Resources Planning Act (RPA) assessments of the status and trends in the US forest resources highlights the difficulties in making accurate projections over time periods of even 20–30 years (S. Prisley, *personal communication*). It is also not clear, for example, whether the expectation for future rates of carbon sequestration should be recalculated each time a new facility with significant feedstock demand comes online. Does the new demand—and its implications for net rate of carbon sequestration in its woodshed going forward—then become part of "BAU"? More generally, the notion of "BAU" seems entirely at odds with the highly dynamic nature of both forest ecosystems and forest economies. The Searchinger et al. (2009) paper is couched in the perspective of the coarse averaging that comes from national- and global-scale assessment of broadly defined sources and sinks for atmospheric carbon. But application of the concept for regional-scale accounting is far more problematic.

It has been pointed out that forest biomass supply is not independent of demand and that landowner responses to strong markets for forest feedstocks can, at least in some regions and under certain conditions, lead to increased supply (Galik et al. 2009; Guo et al. 2011). This would typically result from a combination of bringing more land area into production and implementing methods to increase yield from existing forestland (Abt et al. 2012). Thus, rather than simply extrapolating forward from current rates of sequestration as would be done under a "BAU" baseline, econometric models could be used to project the effect of increasing demand for forest feedstocks on the available supply and stocks of forest carbon. Within some range of market conditions and landowner decisions, greater demand and utilization of forest feedstocks could conceivably lead to greater regional stocks of forest carbon than would be expected under BAU. As a result, while use of a BAU baseline would in most cases be expected to lead to predictions that forest biomass energy production will result in net releases of CO_2 to the atmosphere, incorporation of demand-driven increases in supply (termed "additionality") into the baseline could potentially create the somewhat counterintuitive expectation that forest biomass energy production could lead to greater rates of net carbon sequestration in forests.

Analyses of the potential for these additionality processes have been largely confined to intensive plantation forestry in the southeastern United States (e.g., Abt et al. 2010; Guo et al. 2011). Studies of land-use change in the northeastern United States have highlighted a dramatic increase in forestland over the past century but suggest that total forestland area is likely to have stabilized or even begun to drop (Drummond and Loveland 2010). A larger question in the northeast concerns the fraction of that forestland area that is truly available for harvest, given the wide range of ecological, social, and economic factors that determine whether a landowner is willing to harvest (Butler et al. 2010; Buchholz et al. 2011; Markowski-Lindsay et al. 2012). Outside of plantation forestry, it is rare for forest management practices in the northeast to employ methods such as fertilization or tree planting to boost yield, and it is not clear that the economics of use of forest biomass as energy feedstocks would lead to widespread adoption of such methods in northeastern forests.

Given that differences in the specification of a baseline have such a large impact on the calculation of the net carbon impact of forest biomass energy, and the uncertainty in projecting the future of forest carbon stocks, it shouldn't be a surprise that there is little consensus over the net carbon emissions associated with the use of live woody biomass as a feedstock. It is clear that the use of roundwood as an energy feedstock is not "categorically" carbon neutral, and in regions where there are currently high rates of sequestration of carbon in live forest biomass, policies favoring continued sequestration are likely to have more positive impact on short-term atmospheric carbon than policies favoring harvesting of that forest biomass for use as an energy feedstock. But it is equally clear that rates of sequestration are declining, and a recent study projects that the US forests will cease to be a net sink for carbon within the next 20–30 years (USDA Forest Service 2012). At that point, sustainably produced roundwood feedstocks clearly become even more beneficial in terms of net carbon impact on the atmosphere. But the most fundamental question then becomes whether we can manage a "soft landing" for our forest resources, with equilibration at stable forest carbon stocks, or whether we overshoot supply and deplete carbon stocks regionally through removals that exceed net growth. That issue is addressed in the final section of this chapter.

3.4 Net Carbon Impact of Utilization of Logging Residues

There have been a number of attempts to estimate the availability of logging residues for use as a bioenergy feedstock. For example, Gan and Smith (2006) use 1997 FIA data and an assumption that 70 % of logging residues are recoverable to estimate that as much as 36 million dry tons of residue are available annually at a national level, with most of this in the eastern states. FIA and TPO data suggest that somewhere between 15 and 20 % of the merchantable volume of harvested trees is left onsite as residue, nationally. There are both practical and ecological limits on the fraction of this material that can or should be considered for use as a feedstock, and the Gan and Smith estimate of 70 % utilization rate would be considered high by many studies (Briedis et al. 2011).

Interest in the use of logging residues as an energy feedstock appears to stem from assumptions that it is a source of forest biomass that is underutilized and easily procured (e.g., Manomet Center 2010). Massachusetts has limited the forest feedstocks eligible for renewable energy credits to forest-derived residues, forest-derived thinning, and forest salvage—to the exclusion of roundwood feedstocks—apparently out of concern for potentially unsustainable demand for roundwood feedstocks. The Manomet study that assessed biomass sustainability and carbon policy for Massachusetts concluded that "(t)he removal of tops and limbs generally has little impact on stand level carbon dynamics in Massachusetts forests" (Manomet Center 2010, p 93). But the Manomet study does not directly address the net carbon emissions associated with the use of logging residues as a feedstock.

Field studies that have tested for general impacts of logging on detrital and soil carbon pools in northeastern forests have generated conflicting results (Aber et al. 1978; Binkley 1984; Yanai et al. 2003; Mika and Keeton 2012). Inputs are highly variable in space and time, and mineral soil carbon pools in particular are large relative to annual turnover. As a result, there are significant empirical challenges in measuring impacts of utilization of logging residue on detrital carbon stocks (Yanai et al. 2000). But the science behind detrital carbon dynamics in forests is very clear and leads to the inescapable conclusion that any removal of logging residue from a stand for use as a feedstock will reduce the stock of detrital carbon at the site over time. While analyses of forest carbon dynamics differ widely in the level of complexity with which they treat decomposition and the fate of organic matter, they generally all characterize decomposition as a "donor-controlled" process (Harmon et al. 1986; Gholz et al. 2000; Yanai et al. 2003). Removal of some fraction of logging debris following a harvest will necessarily result in a reduction in the mass of that pulse of carbon that remains on the site, averaged over any subsequent time period. And for time periods of less than 50 years, in which per capita (exponential) decay rates are typically assumed to remain relatively constant, the relative magnitude of the reduction in the average mass of the pulse remaining on the site will be directly proportional to the fraction of the residue that was removed.

In the purely hypothetical case of a forest at steady state with constant annual inputs of litter, it is a simple matter to calculate the expected mass of the forest floor with and without removal of some fraction of the residue inputs, as a function of forest type/region-specific decay rates (as used, for example, in Smith and Heath 2002). In this case, the expected forest floor mass is approximately equal to *inputs/decay rate* (where *decay rate* is in units of fraction lost per unit time). A more precise estimate would typically be generated using a discrete time step simulation of annual litter inputs and losses. But the general principle is clear and appears to be widely accepted: any reduction in the magnitude of the litter input will result in a proportional reduction in the mass of the detrital carbon stocks on the site.

The same principle applies to the determination of the fate of a pulse of logging residue generated by a harvest. If some fraction of that pulse is removed for use as a feedstock, there will be a proportional reduction in the mass of the pulse remaining on the site over time. The actual mass remaining at any given time will decline exponentially as a function of the characteristic decay rate for the climate and species at the site. The fact that the amount remaining after 100 years is expected to be only several percent of the original mass of the pulse (and reflects the transfer of material to very-long-term mineral soil carbon pools) has led some to argue that since most of the material was going to decompose "anyway," combustion of the material as a feedstock should constitute an effectively carbon-neutral energy source. The logical flaw in this argument is that over policy-relevant time scales, the mass of carbon sequestered in a forest as a result of the pulse of logging residue is not trivial (Fig. 3.2). For northeastern forests, decay rates for logging residue are expected to be in the range from 5 to 10 % per year (Smith and Heath 2002). But even given these relatively high rates of decay, the average mass of the pulse of residue remaining over a 30-year period would be 30–50% of the original mass of the residue (Fig. 3.2). Thus, every 100 T of logging residue used as a feedstock would be expected to reduce the average mass of detrital carbon at the site (and in the region as a whole) by 30–50 T when averaged over a 30-year time period. In contrast to roundwood feedstocks, logging residue does not "regrow" per se, and this reduction in the average mass of detrital carbon at a site represents a net release of carbon to the atmosphere. While the argument presented here is based on simple principles of mass balance, much more detailed analyses using ecosystem models reach similar conclusions (Schlamadinger et al. 1995; Palosuo et al. 2008; Repo et al. 2011; Domke et al. 2012).

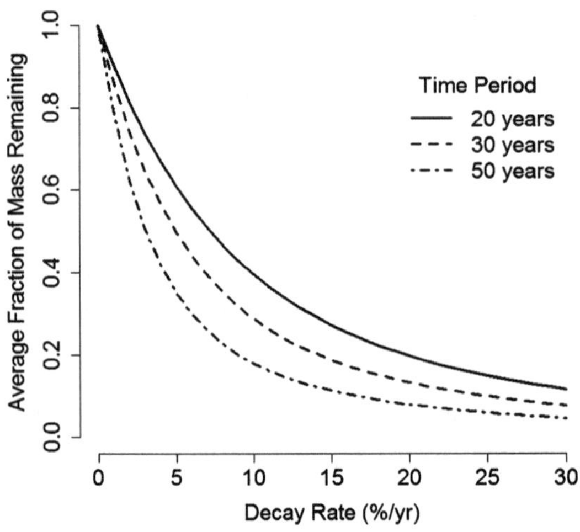

Fig. 3.2 Fraction of the mass of an initial input of logging residue remaining on site, averaged over time periods ranging from 20 to 50 years, as a function of variation in annual decay rates. Use of 1 Mg of logging residue as a feedstock would therefore reduce the average forest floor mass in the region and forest type from which the feedstock was sourced, over the defined time period, by this amount. Forest floor decay rates from Smith and Heath (2002) range from 4 to 26 % per year for different regions and forest types of the United States (outside of Alaska)

3.5 Landscape-Scale Determinants of Biomass Supply in Northeastern Forests

In the simplest terms, the overall "supply" of biomass in a region (i.e., net increment available for harvest) can be seen as the product of two basic relationships: (1) the relationship between stand biomass and biomass growth rate (increment) and (2) the variation in stand biomass (tons/hectare) in a region (Buchholz et al. 2011). The annual growth of a forest stand (i.e., the annual increment in biomass) varies predictably as a function of total stand biomass (in addition to many other variables). In principle, growth is expected to increase as stands increase in total biomass (because of the inherent scaling of tree growth to tree size) and then reach a peak at some intermediate biomass. As stands continue to increase in biomass beyond that point, yields should decline and eventually reach zero. The reasons for the decline have been widely debated in the ecological literature, but the fact that the decline happens is indisputable—forests do not continue to increase in living biomass indefinitely. The shape of this biomass increment curve (Fig. 3.3a) is central to the overall yield of a landscape or a region, when combined with the information on the proportion of the landscape in stands of different total biomass (Fig. 3.3b) (Buchholz et al. 2011).

Figure 3.3 highlights differences across the northeastern United States in (1) the magnitude of the peak biomass increment, (2) the level of total tree biomass at

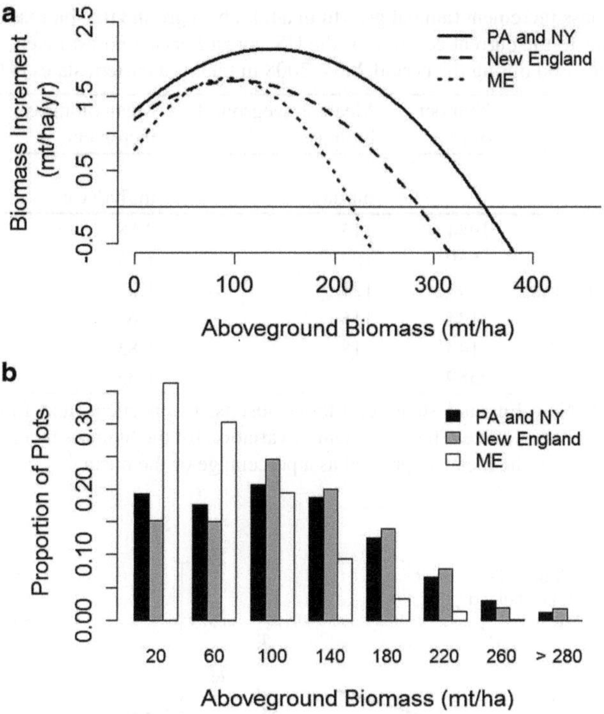

Fig. 3.3 (**a**) Plot-level aboveground biomass increment as a function of aboveground biomass for three regions of the northeastern United States: (1) Pennsylvania and New York; (2) the New England states of Vermont, New Hampshire, Massachusetts, Connecticut, and Rhode Island; and (3) Maine. The *curves* are maximum likelihood fits of a quadratic function to unharvested FIA plots recensused between 2004 and 2008. (**b**) The frequency distribution of FIA plots (harvested and unharvested) for those three regions, as a function of aboveground biomass. Redrawn from Buchholz et al. (2011)

which the peak occurs, and (3) the biomass at which increment is predicted to become zero. The New England states have a lower peak biomass increment than Pennsylvania and New York, and the peak occurs at lower total tree biomass. The lower yields at any given tree biomass presumably reflect a combination of the effects of less favorable climate and soils. As a result, expected maximum stand biomass (i.e., the biomass at which stands are assumed to, on average, stop accumulating biomass in the aboveground portions of live trees) varies significantly across the region, from slightly over 200 mt/ha in Maine to over 350 mt/ha in Pennsylvania and New York.

The average biomass increment (growth per hectare) averaged across an entire region (Table 3.1) is a product of the biomass increment functions described above and the stand-level variation in aboveground tree biomass in the region (Fig. 3.3b). Despite what appear to be common assumptions that northeastern forests overall are nearing "maturity" (and zero net aboveground live biomass increment)

Table 3.1 Biomass increment (annual growth in adult aboveground tree biomass, over the period from the previous to the current census) for the US Forest Service Forest Inventory and Analysis (FIA) plots recensused during the period 2004–2008 in the northeastern states (FIADB 4.0)

	Number of plots	Mean aboveground biomass	Mean biomass increment	Coefficient of variation
				% of biomass
State		mt/ha	mt/ha/year	increment (%)
Pennsylvania	1944	115	2.08	3.5
New York	559	103	1.23	9.5
Southern New England	191	121	1.07	20.2
Vermont	134	115	1.64	12.7
New Hampshire	143	118	1.83	10.3
Maine	2589	68	1.35	3.0

The southeastern New England states of Massachusetts, Connecticut, and Rhode Island were lumped to increase sample size. The coefficient of variation for the biomass increment is computed as the standard error of the mean, expressed as a percentage of the mean

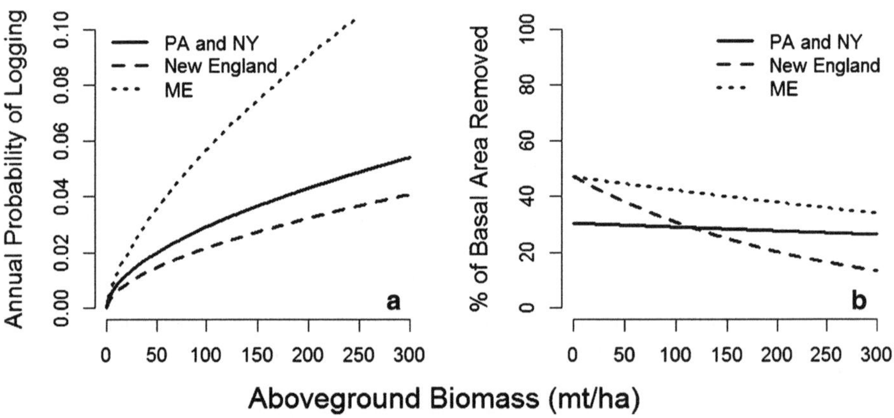

Fig. 3.4 Harvest regimes for three regions of the northeastern United States (using regions defined in Fig. 3.3). Partial harvesting is common in all three regions. The harvest regimes were derived from FIA data using statistical methods outlined in Canham et al. 2013. (**a**) The annual probability that a plot is logged, as a function of aboveground biomass. (**b**) The percent of tree basal area removed (cut down), as a function of aboveground biomass, given that a plot was logged. The *curves* represent the means of the functions, but there is enormous variability in the percentage of basal area removed in individual plots (Canham et al. 2013)

(Siccama et al. 2007), forest inventory data show a landscape with an extremely wide range of biomass, with most stands clustered in the range of adult tree biomass that has the highest rates of biomass increment (Canham et al. 2013). As would be expected from the high rates of harvest and cold climate in Maine (Fig. 3.4), the frequency distribution for that state is dominated by stands with very low biomass, and the mean biomass is dramatically lower than in the other states (Table 3.1). The statewide average biomass increment is also relatively low (1.35 mt/ha/yr).

The southeastern New England states (Massachusetts, Connecticut, and Rhode Island) have the highest average biomass, with the greatest fraction of plots in high biomass stands, but, partly as a result of this, have the lowest average biomass increment (Table 3.1). Pennsylvania has intermediate average biomass (115 mt/ha on average) but the highest average biomass increment (2.08 mt/ha/yr, Table 3.1). This presumably reflects favorable soils and climates (relative to the other northeastern states) but also reflects a distribution of plot biomass that is concentrated in the range of biomass where biomass increment is at its peak (Fig. 3.3). Given that the peak predicted biomass increment is only slightly higher than the current average biomass increment, the current frequency distribution of plot-level biomass in Pennsylvania (Fig. 3.3b) appears to be near-optimal for biomass increment.

3.6 Regional Harvest Regimes: Managing a "Soft Landing" for Northeastern Forest Resources

Regional harvest levels fluctuate on time scales faster than forested landscapes can equilibrate. As a result, the stand biomass distributions shown in Fig. 3.3b are not expected to be in steady state with the current forest harvest regimes (Canham et al. 2013). The frequency distribution of stand biomass—and the resulting net biomass increment of the landscape as a whole—in each of these regions will gradually change over time as a result of the combination of (1) changes in factors that affect the recruitment, growth, and mortality of trees in a stand and (2) the regional harvest regime. The history of forest resources in the United States over the past 200 years has been marked by dramatic fluctuations in both the area and average biomass of forestland within any given region (Drummond and Loveland 2010). While there are a variety of laws, regulations, and certification schemes designed to ensure sustainable forest management, there are large segments of the working forestland base that are not covered by laws that would guarantee that forests are not over-harvested in the future. Given that most analyses conclude that forest carbon stocks within the northeast and for the United States as a whole will peak at some point in the near future (Hurtt et al. 2002; USDA Forest Service 2012), the most basic question would appear to be whether we can achieve a smooth transition to a forest landscape that stabilizes at some level of carbon stocks while providing a long-term sustainable supply of biomass for both traditional forest products and bioenergy production.

There has been a sea change in management of northeastern forests over the past 50 years, and clear-cutting has become relatively rare outside of conifer-dominated ecosystems (Birdsey and Lewis 2003; Smith et al. 2009). Instead, a wide variety of silvicultural systems using partial harvesting have become common, particularly in hardwood-dominated forests. While individual harvests are almost infinitely variable, the aggregate harvest disturbance "regime" in a region can be characterized statistically in terms of overall disturbance frequency, intensity and the selectivity of the mortality as a function of tree species and sizes. Canham et al. (2013) have developed simple statistical models to characterize regional harvest regimes, using

FIA data (Fig. 3.4). Their analyses confirm that relatively frequent partial harvests are the most prevalent silvicultural system in the northeastern United States, but with enormous variability across regions and forest types, and in any particular stand (Fig. 3.4). Their analyses quantify three different aspects of aggregate harvest regimes: (1) the annual probability that a plot is logged, as a function of total aboveground tree biomass; (2) the fraction of tree basal area removed, if a plot was logged, again as a function of total aboveground tree biomass; and (3) the probability that an individual tree within a plot was removed, if the plot was logged, as a function of the fraction of basal area removed at the plot level, the species of tree, and its size (DBH).

Failure to appreciate the regional impact of the logging regime can lead to misleading assumptions about the carbon dynamics of forested landscapes. For example, the expected decline in net rates of carbon sequestration in northeastern forests is assumed to be partly a function of the maturing of a largely "even-aged" forest landscape that dates from clear-cutting or land abandonment more than a century ago (Hurtt et al. 2002; Siccama et al. 2007). However, any forester in the Northeast can attest that the vast majority of forestland in the Northeast is actively managed in one way or another. Recent analyses suggest that harvesting has an enormous impact on the magnitude of the carbon sink in the US forests (Zheng et al. 2011). Moreover, harvesting is not just an important source of adult tree mortality; it is currently a larger source of tree mortality in northeastern forests than all other causes of mortality (natural and anthropogenic) combined (statistics from FIA Web site, summarized in Canham et al. 2013). Thus, it would be erroneous to conclude that much of the forest landscape in the northeastern United States has reached a successional stage and level of aboveground biomass where the potential to sequester carbon has disappeared (Fig. 3.3b).

These harvest regimes have a more prosaic implication for the way that ecologists think about northeastern forests. Stand "age" is one of our most cherished metrics of forest ecosystem status. The results of Canham et al. (2013) suggest that except for the small fraction of the landscape that has been reserved from harvest over the past century, stand age will be difficult to assign as a single, predictive metric of forest ecosystem status (*sensu* Odum 1969). The patterns of logging illustrated in Fig. 3.2 imply that stands will be partially harvested at least several times over the course of a century, with a significant fraction of aboveground tree biomass removed in any given harvest.

The harvest regimes in Fig. 3.4 can be combined with the relationships in Fig. 3.3 to run a simple stochastic simulation model that estimates the hypothetical long-term and steady-state consequences of different harvest regimes, given the current frequency distribution of stand biomass (Fig. 3.3b) and the characteristic biomass increment curve (Fig. 3.3a) for a region. The simulations use the current frequency distribution of stand biomass to populate a hypothetical landscape of 20,000 "stands." Changes in the biomass of those individual stands over time are then a stochastic function of (1) harvests, applied stochastically as a function of the mean and variances of the relationships in Fig. 3.4, and (2) the biomass increment of the stand, again as a stochastic function of the mean and variance of the relationships

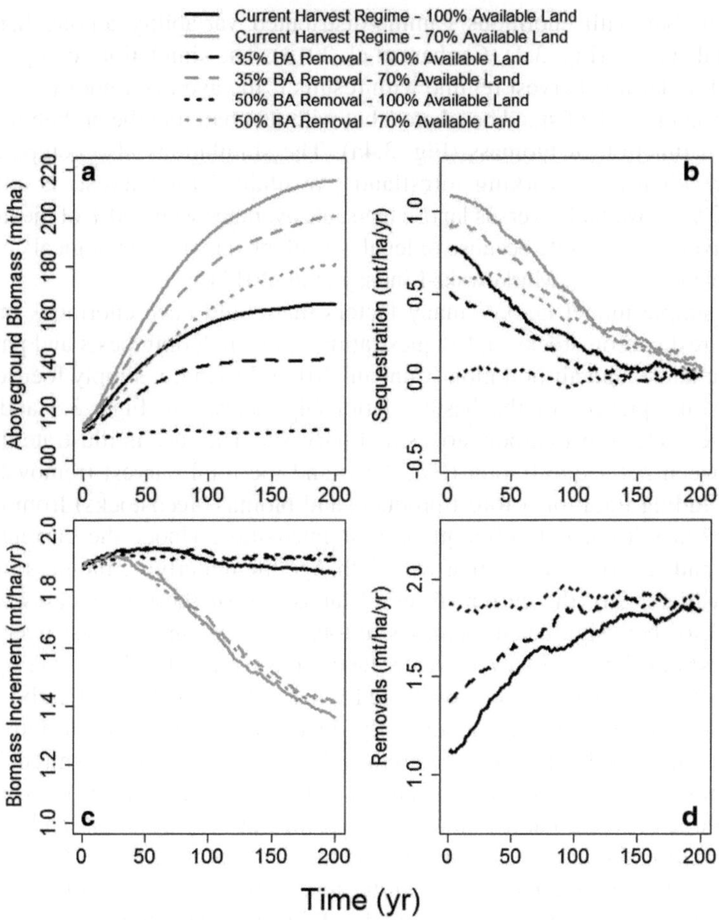

Fig. 3.5 Results of a simple stochastic simulation model projecting average landscape-level changes for the states of Pennsylvania and New York in (**a**) aboveground woody biomass, (**b**) net annual change in aboveground woody biomass (sequestration), (**c**) biomass increment (net growth without consideration of removals), and (**d**) harvest (removals). The simulations use the biomass increment functions from Fig. 3.3a and the harvest regimes from Fig. 3.4 for PA and NY. Three variants of the harvest regime for those states were simulated: the current harvest regime (~30 % of basal area removed from a stand, when harvested, with gamma-distributed variability among individual stands) and hypothetical regimes in which average harvest intensity is increased to either 35 or 50 % of stand basal area. Two variants of each of the three regimes are illustrated: one in which results are averaged over only the available "working" forestland, and one in which 30 % of the landscape is assumed to be effectively reserved and unavailable for harvest

shown in Fig. 3.3a. Average landscape-level metrics are then simply averaged across the individual stands.

Figure 3.5 illustrates the hypothetical long-term consequences of six different harvest regimes for the states of Pennsylvania and New York. The current harvest regime in that region removes on average 30 % of the basal area when a stand is harvested (but with enormous gamma-distributed variability among harvests in individual stands) (Fig. 3.4) (Canham et al. 2013). The simulations compare results expected under this harvest regime with results if the average removal is increased to either 35 or 50 % of stand basal area, but without changing the probability of logging as a function of biomass (Fig. 3.4a). The simulations also compare results expected when only working forestland ("available" for harvest) is considered (i.e., 100 % is available) versus landscape-scale averages when 30% of the landscape is effectively unavailable because of legal, social, physical, or ecological constraints (Buchholz et al. 2011; Markowski-Lindsay et al. 2012).

This simple model ignores many factors that could have enormous impact on future forest carbon stocks and sequestration rates, including pests and pathogens, climate change, and air pollution (Canham 2013). Instead, it simply focuses on the logical consequences of the basic relationships shown in Figs. 3.3 and 3.4 for landscape-scale forest carbon stocks and harvests. The results illustrate the trade-offs between net sequestration (Fig. 3.5b) and the total harvest (removals for all uses, including traditional forest products and biomass feedstocks) from the landscape (Fig. 3.5d) under different harvest intensities. Under the current harvest regime (and ignoring factors that would change the underlying biomass increment curve in Fig. 3.3a), the current forested landscape of these two states would be expected to show significant increases in total aboveground biomass over the next 100 years (Fig. 3.5a), with even greater landscape average levels if some significant fraction of the forestland is not available for harvest. The net rate of carbon sequestration would be expected to gradually decrease to zero as total aboveground biomass stabilizes, but total harvest (removals) would show a concomitant increase. At that point, biomass increment would be balanced by removals (Fig. 3.5c, d).

The three different harvest intensities stabilize at very different landscape average aboveground biomass carbon stocks (Fig. 3.5a), ranging from 150 mt/ha under the current harvest regime to 110 mt/ha under the 50 % BA removal scenarios (essentially no change from current stocks, Table 3.1). The three harvest intensities stabilize in roughly 200 years at very similar average rates of biomass increment and removals (~1.9 mt/ha/yr), and the most intensive harvest regime (50 % average BA removal) achieves that rate throughout the period. But the sum of average sequestration and average removals (in metric tons/ha/yr) for the three different intensities of harvest is almost identical over the entire time period. In effect, more intensive harvest regimes produce correspondingly greater biomass removals for use in either traditional timber products or as bioenergy feedstocks, but they do so at the expense of reducing landscape average aboveground carbon stocks, with almost a 1:1 trade-off.

These trade-offs would seem to have important policy implications. The current harvest regimes would allow a gradual but significant increase in total forest carbon stocks per unit area of forestland, with even greater increases when averaged over both the working forests and the effectively "reserved" (unavailable) forests (Fig. 3.5a). More intensive harvest regimes, in contrast, would provide a higher annual supply of biomass, some fraction of which could be used as a bioenergy feedstock. The differences in net sequestration versus additional potential feedstock supply appear to be roughly equal in the magnitudes of the landscape-scale carbon fluxes involved, and the different harvest regimes eventually equilibrate at roughly equal sustainable harvest volume. So implementing higher harvest regimes now does not appear to produce a "soft landing" to a landscape configuration with a higher or a lower long-term sustainable supply.

But the cumulative net carbon impacts of the different harvest intensities are very different. Because of differences in energy density and efficiencies in conversion of biomass feedstock to useable energy, the carbon in one metric ton of a forest feedstock will in most cases offset less than one metric ton of carbon emissions from a fossil fuel feedstock (Buchholz et al. 2011). In contrast, allowing that metric ton of forest carbon to be sequestered provides a direct 1:1 offset to a fossil fuel emission. The overall environmental benefits of sequestration are likely to be even greater, since these forests provide a wide range of ecosystem services, and those services can be expected to be positively related to increases in landscape average forest biomass (Nelson et al. 2009).

This simple model focuses only on net carbon impacts and ignores the myriad ecological, social, and economic consequences of different intensities of regional harvest regimes. The model is also inherently "equilibrial" and illustrates the path to steady state that would occur under the highly simplified structure of the model and the consequences of a specific regional harvest regime. But there is little—either in theory or in our history—to suggest that the dynamics of either regional forest landscapes or regional forest economies are inherently steady state. The fraction of the forestland base in the northeastern United States managed under legal or voluntary sustainable management schemes (including various forest certification programs) has increased in the past 20 years (Espinoza et al. 2012), but traditional definitions of sustainable forestry, even those embodied in certification schemes such as those of the Sustainable Forest Initiative and the Forest Stewardship Council, do not necessarily ensure low net carbon emissions from forest biomass energy over policy-relevant time scales.

3.7 Conclusions

There has been enormous debate about the net impact on atmospheric carbon from the use of forest feedstocks for energy production. There appears to be little consensus over even the methodology for calculating net carbon emissions associated with biomass energy. Use of different "baselines" leads to very different conclusions about the benefits of the use of roundwood feedstocks. Careful consideration of the dynamics of forest carbon cycling can lead to conclusions that run counter to conventional wisdom. For example, there has been widespread interest in the use of logging residue as an energy feedstock, presumably because of the assumption that this is an "unused" or a "waste" resource that should be captured for energy production, rather than simply left to decompose in the forest. But as pointed out earlier in this chapter, any widespread utilization of logging residue can be expected to reduce regional detrital carbon stocks relative to traditional forest management practices where that residue was left in the forest. And that reduction in regional detrital carbon stocks would represent a significant net release of carbon to the atmosphere.

While much of the debate about forest biomass energy in the northeastern United States has focused on the largely methodological issues of assessing net carbon emissions of different combinations of forest feedstocks and energy production technologies (i.e., Milbrandt 2008; Sherman 2007), the more critical issues appear to be the assessment of the magnitude of the sustainable supply of forest biomass and the future of that resource in an uncertain environment. Since most analyses predict that rates of carbon sequestration in northeastern forests will decline over the next 50 years, it seems likely that the relative benefits of policies favoring sequestration versus intensification of harvest will shift over the next 50 years in favor of increased biomass energy production. But it is equally clear that this will only be the case if some combination of ecological, social, economic, and political forces ensure that harvest rates are sustainable and do not deplete total forest carbon stocks regionally.

References

Abbas D, Current D, Phillips M, Rossman R, Hoganson H, Brooks KN (2011) Guidelines for harvesting forest biomass for energy: a synthesis of environmental considerations. Biomass Bioenerg 35:4538–4546

Aber JD, Botkin DB, Melillo JM (1978) Predicting effects of different harvesting regimes on forest floor dynamics in northern hardwoods. Can J For Res 8:306–315

Abt KL, Abt RC, Galik C (2012) Effect of bioenergy demands and supply response on markets, carbon, and land use. Forest Sci 58:523–539

Abt RC, Abt KL, Cubbage FW, Henderson JD (2010) Effect of policy-based bioenergy demand on southern timber markets: a case study of North Carolina. Biomass Bioenerg 34:1679–1686

Binkley D (1984) Does forest removal increase rates of decomposition and nitrogen release. Forest Ecol Manag 8:229–233

Birdsey RA, Lewis GM (2003) Current and historical trends in use, management, and disturbance of U.S. forestlands. In: Kimble JM et al (eds) The potential of US forest soils to sequester carbon and mitigate the greenhouse effect. CRC Press, New York, pp 15–33

Briedis JI, Wilson JS, Benjamin JG, Wagner RG (2011) Logging residue volumes and characteristics following integrated roundwood and energy-wood whole-tree harvesting in central Maine. North J Appl For 28:66–71

Buchholz T, Canham CD, Hamburg SP (2011) Forest biomass and bioenergy: opportunities and constraints in the Northeast United States. Report issued by the Cary Institute of Ecosystem Studies, Millbrook, NY, http://www.ecostudies.org/report_biomass_2011.pdf

Butler BJ, Ma Z, Kittredge DB, Catanzaro P (2010) Social versus biophysical availability of wood in the northern United States. North J Appl For 27:151–159

Canham CD. Disequilibrium and transient dynamics: disentangling responses to climate change versus broader anthropogenic impacts on temperate forests of eastern North America. In: Burslem D, Coomes D (eds) Forests and global change. Oxford University Press, London (in press)

Canham CD, Pace ML, Papaik MJ, Primack AGB, Roy KM, Maranger RJ, Curran RP, Spada DM (2004) A spatially explicit watershed-scale analysis of dissolved organic carbon in Adirondack lakes. Ecol Appl 14:839–854

Canham CD, Rogers N, Buchholz T (2013) Regional variation in forest harvest regimes in the northeastern United States. Ecol Appl 23:515–522

Cook J, Beyea J (2000) Bioenergy in the United States: progress and possibilities. Biomass Bioenerg 18:441–455

Domke GM, Becker DR, D'Amato AW, Ek AR, Woodall CW (2012) Carbon emissions associated with the procurement and utilization of forest harvest residues for energy, northern Minnesota, USA. Biomass Bioenerg 36:141–150

Drummond MA, Loveland TR (2010) Land-use pressure and a transition to forest-cover loss in the eastern United States. Bioscience 60:286–298

Espinoza O, Buehlmann U, Smith B (2012) Forest certification and green building standards: overview and use in the U.S. hardwood industry. J Clean Prod 33:30–41

Galik CS, Abt R, Wu Y (2009) Forest biomass supply in the Southeastern United States – implications for industrial roundwood and bioenergy production. J Forest 107:69–77

Gan JB, Smith CT (2006) Availability of logging residues and potential for electricity production and carbon displacement in the USA. Biomass Bioenerg 30:1011–1020

Gholz HL, Wedin DA, Smitherman SM, Harmon ME, Parton WJ (2000) Long-term dynamics of pine and hardwood litter in contrasting environments: toward a global model of decomposition. Glob Change Biol 6:751–765

Guo ZM, Hodges DG, Abt RC (2011) Forest biomass supply for bioenergy production and its impacts on roundwood markets in Tennessee. South J Appl Forest 35:80–86

Harmon ME, Franklin JF, Swanson FJ, Sollins P, Gregory SV, Lattin JD, Anderson NH, Cline SP, Aumen NG, Sedell JR, Lienkaemper GW, Cromack K, Cummins KW (1986) Ecology of coarse woody debris in temperate ecosystems. Adv Ecol Res 15:133–302

Hurtt GC, Pacala SW, Moorcroft PR, Caspersen J, Shevliakova E, Houghton RA, Moore B (2002) Projecting the future of the US carbon sink. Proc Natl Acad Sci U S A 99:1389–1394

Manomet Center for Conservation Sciences (2010). Biomass sustainability and carbon policy study. Prepared for Commonwealth of Massachusetts, Department of Energy Resources, p 182

Markowski-Lindsay M, Catanzaro P, Damery D, Kittredge DB, Butler BJ, Stevens T (2012) Forest-based biomass supply in Massachusetts: how much is there and how much is available. J Environ Manag 106:1–7

Mika AM, Keeton WS (2012) Factors contributing to carbon fluxes from bioenergy harvests in the U.S. Northeast: an analysis using field data. Glob Change Biol Bioenergy 5(3):290–305

Milbrandt A (2008) A geographic perspective on the current biomass resource availability in the United States. Technical report, NREL/TP-560-39181 (2008 update), p 70

Nelson E, Mendoza G, Regetz J, Polasky S, Tallis H, Cameron DR, Chan KMA, Daily GC, Goldstein J, Kareiva PM, Lonsdorf E, Naidoo R, Ricketts TH, Shaw MR (2009) Modeling multiple ecosystem services, biodiversity conservation, commodity production, and tradeoffs at landscape scales. Front Ecol Environ 7:4–11

Odum EP (1969) The strategy of ecosystem development. Science 164:262–270

Palosuo T, Peltoniemi M, Mikhailov A, Komarov A, Faubert P, Thurig E, Lindner M (2008) Projecting effects of intensified biomass extraction with alternative modelling approaches. Forest Ecol Manag 255:1423–1433

Radeloff VC, Mladenoff DJ, Gustafson EJ, Scheller RM, Zollner PA, He HS, Akcakaya HR (2006) Modeling forest harvesting effects on landscape pattern in the Northwest Wisconsin Pine Barrens. Forest Ecol Manag 236:113–126

Repo A, Tuomi M, Liski J (2011) Indirect carbon dioxide emissions from producing bioenergy from forest harvest residues. Glob Change Biol Bioenergy 3:107–115

Schlamadinger B, Spitzer J, Kohlmaier GH, Ludeke M (1995) Carbon balance of bioenergy from logging residues. Biomass Bioenerg 8:221–234

Schulze E-D, Korner C, Law BE, Haberl H, Luyssaert S (2012) Large-scale bioenergy from additional harvest of forest biomass is neither sustainable nor greenhouse gas neutral. Glob Change Biol Bioenergy. doi:10.1111/j.1757-1707.2012.01169.x

Searchinger T, Hamburg S, Melillo J, Chameides W et al (2009) Fixing a critical climate accounting error. Science 326

Sherman AR (2007) Vermont wood fuel supply study – an examination of the availability and reliability of wood fuel for biomass energy in Vermont. Biomass Energy Resource Center, Montpelier, VT, p 76

Siccama TG, Fahey TJ, Johnson CE, Sherry TW, Denny EG, Girdler EB, Likens GE, Schwarz PA (2007) Population and biomass dynamics of trees in a northern hardwood forest at Hubbard Brook. Can J For Res 37:737–749

Smith JE, Heath LS (2002) A model of forest floor carbon mass for United States forest types. USDA Forest Service Northeastern Research Station research paper NE-722

Smith WB, Miles PD, Perry CH, Pugh SA (2009). Forest resources of the United States, 2007. General technical report WO-78. U.S. Department of Agriculture, Forest Service, p 336

USDA Forest Service (2012) The future of America's forests and rangelands – The 2010 Resources Planning Act (RPA) assessment. USDA, Washington, DC, in press

U.S. Environmental Protection Agency (2011) Accounting framework for biogenic CO_2 emissions from stationary sources. US Environmental Protection Agency, Office of Atmospheric Programs, Climate Change Division, Washington DC, p 140, http://www.epa.gov/climatechange/emissions/biogenic_emissions.html

Yanai RD, Arthur MA, Siccama TG, Federer CA (2000) Challenges of measuring forest floor organic matter dynamics: repeated measures from a chronosequence. Forest Ecol Manag 138:273–283

Yanai RD, Stehman SV, Arthur MA, Prescott CE, Friedland AJ, Siccama TG, Binkley D (2003) Detecting change in forest floor carbon. Soil Sci Soc Am J 67:1583–1593

Zanchi G, Pena N, Bird N (2012) Is woody bioenergy carbon neutral? A comparative assessment of emissions from consumption of woody bioenergy and fossil fuel. Glob Change Biol Bioenergy 4:761–772

Zheng DL, Heath LS, Ducey MJ, Smith JE (2011) Carbon changes in conterminous US forests associated with growth and major disturbances: 1992–2001. Environ Res Lett 6:014012

Chapter 4
Traditional Breeding, Genomics-Assisted Breeding, and Biotechnological Modification of Forest Trees and Short Rotation Woody Crops

Joshua R. Herr and John E. Carlson

4.1 Introduction

The use of woody biomass for biofuels is being pursued with both targeted harvests from natural forests and growth of Short Rotation Woody Crops (SRWCs) in plantations. Both native and exotic tree species can contribute to energy feedstocks, as managed, unmanaged, or native forests. However, although standing woody biomass may in total be considered sufficient to meet projected demands for energy in certain Northern regions, increases in productivity are often still required for woody biomass to become an economically and ecologically sustainable source of energy. Traditional plant breeding has been very successful in improving growth, tree volume, and various wood quality traits for the lumber and paper industries (Harfouche et al. 2012) for forest trees and for SRWCs. The rate of progress in tree breeding is limited by the long breeding cycle times of most tree species, however. Biotechnological approaches have great potential to augment and help advance tree improvement programs, through early, indirect selection of improved genotypes, propagation through tissue culture, or genetic engineering of traits such as flowering time or wood quality.

The extensive Northern natural forests currently represent the major source of readily available woody biomass for renewable energy. Targeted harvesting of "weedy" species and small diameter trees through pre-commercial thinning of natural stands can provide large amounts of biomass while also improving round-timber productivity. This practice can be sustainable as shown by many generations of harvests of small diameter trees previously as pulp wood. In this scenario, tree genetics can play a role through improvement of seed sources for artificial regeneration or through the use of genetic markers to identify the best trees to leave as parents for

J.R. Herr, Ph.D. (✉) • J.E. Carlson, Ph.D.
The Department of Ecosystem Science and Management, The Schatz Center for Tree Genetics, Pennsylvania State University, 321 Forest Resources Building, University Park, PA 16802, USA
e-mail: jherr@psu.edu; jec16@psu.edu

M. Jacobson and D. Ciolkosz (eds.), *Wood-Based Energy in the Northern Forests*, DOI 10.1007/978-1-4614-9478-2_4, © Springer Science+Business Media New York 2013 the next generation when selective harvesting is practiced. Tree improvement for lumber, pulp, and bioenergy focus on many of the same traits including growth, yield, and wood properties as the main priorities, along with other traits such as disease resistance and resistance or tolerance to abiotic stresses (Neale and Kremer 2011). Thus genetic improvement for lumber production will also directly benefit yields of the targeted biomass harvests during tree thinning in managed forest stands. Tree breeding uses recurrent selection at the population level to gradually improve productivity of forest species (Neale and Kremer 2011). This helps to maintain genetic diversity as a buffer against future changing conditions, thus supporting sustainability for all forest products, including bioenergy, as well as ecological services.

Interest is also increasing in the use of dedicated biomass energy plantations because of attributes such as uniformity of production and ease of harvesting relative to natural forests. Plantations may indirectly help to maintain biodiversity by reducing pressure on natural forests, by mitigating climate change through enhanced carbon-sequestration, and by providing a wider array of commercial value-added chemicals and products (Harfouche et al. 2011). Plantation management of trees can also help ensure reliability of biomass supply to energy producers and to supplement fiber from forests.

SRWCs are a valuable renewable resource with the ability to provide both combustion heat and liquid biofuels. With a long history as a feedstock for the pulp and paper industry, SRWCs are now being touted as a component of future plant-based bioenergy allocations by world governments. Fast-growing trees with high productivity, such as poplar (*Populus* species and hybrids) and willow (*Salix* species and hybrids) have been deemed the most productive SRWCs in temperate climates, while Pines (*Pinus* species) and *Eucalyptus* (both species and hybrids) are perhaps most advantageous in seasonally warmer areas (Neale 2007; Neale and Kremer 2011). There are many benefits of SRWCs as opposed to other bioenergy crops: flexible year-round harvesting, low-input multiple-growing season growth, reduced storage costs, and reduced storage degradation losses are a few of these benefits (Sims and Venturi 2004). Additionally, SRWC plantations can help to subvert climate change by contributing to carbon sequestration, replacing fossil fuels with renewable wood bioenergy crops, and by supporting retention of lands in plant biomass production and away from urban development (Neale and Kremer 2011).

In the future, increasing demand for both high quality sources of bioenergy and forest products will create large global productivity targets for renewable fiber production (Herr 2011) as evidenced by research over the last two decades (Fig. 4.1). Current targets for SRWCs, such as poplar and willow, are 18-22 dry tonnes per hectare per year (Cobb et al. 2008). Eastern cottonwood (*Populus deltoides*) produces an annual average of 11-18 dry tonnes per hectare (Simmons et al. 2008). Early research in short rotation willow produced site-specific yields in the range of 7-16 dry tonnes per hectare per year (Mead 2005). A 20-year rotation cycle of Loblolly Pine (*Pinus taeda*) produced a yearly harvest of 9 dry tonnes per hectare (Simmons et al. 2008). Current dry mass production levels for forest trees and SRWCs are not sufficient to meet future demand for plant-based bioenergy—and as a result, improvements of woody feedstocks are needed.

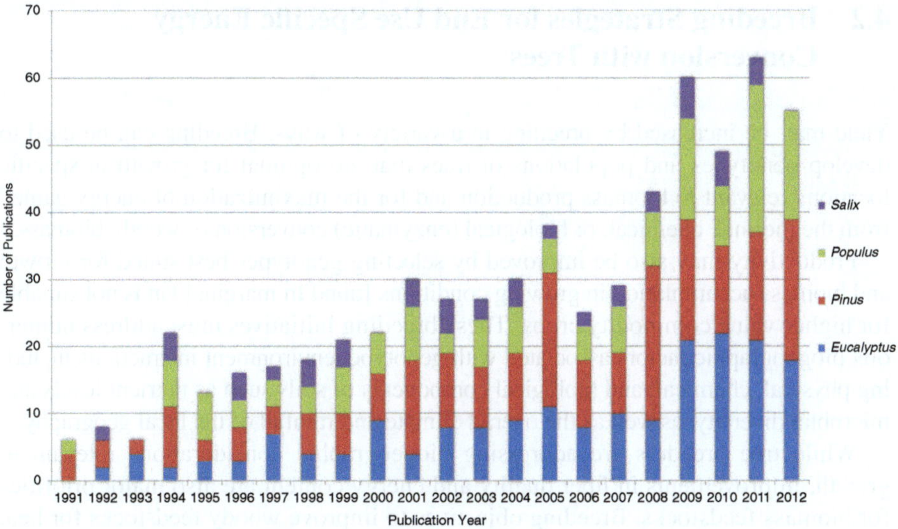

Fig. 4.1 Graph showing the number of published papers over the last 20 years that include *both* a SRWC species name (*Eucalyptus*, *Populus*, *Pinus*, and *Salix*) and the words "plant breeding" in the title, abstract, or keywords

In order to avoid competition with food crops, SRWCs are typically grown on marginal lands that may not support high productivity crop growth. As a result, to expand the range of SRWC production, it will be imperative for breeding to improve efficient water use and nutrient acquisition to be productive on these marginal sites. Additionally, marginal sites are typically associated with increased environmental stresses on plants, from both abiotic and biotic sources. Breeding resources can be developed to address susceptibility to frequently encountered stresses. With the added benefit of optimization for additional end uses (such as chemical co-production, pulp and paper production, climate change mitigation efforts, strip mine reclamation, etc.), genetic improvement of trees for biomass has the potential for increased growth and biomass yields and adaptation to marginal lands and environments (Mizrachi et al. 2012). Breeding can also accomplish reduction of inputs such as irrigation, fertilizers, and pesticides, along with improved downstream resources for the bioenergy industry, such as enzymes used to digest plant material, catalysts for improved saccharification, and improved thermal and chemical pretreatments (Mansfield 2009).

For the development of specific hybrids and cultivars, breeding pairs for SRWC are typically selected on projected biomass yields, wood quality, rapid growth at close spacing, the ability to be efficiently vegetatively propagated, and traits associated with environmental stress tolerance, such as drought or herbivore tolerance and resistance to various pests and diseases (Cobb et al. 2008), while breeding with most forest trees relies on recurrent selection on an aggregate value of multiple traits simultaneously with long-term tests used for gradual improvement of populations rather than specific genotypes.

4.2 Breeding Strategies for End Use Specific Energy Conversion with Trees

Yield may be increased by breeding in a variety of ways. Breeding can be used to develop genotypes and populations of trees that are optimal for growth in specific locations relevant to biomass production and for the maximization of energy gained from the thermal, chemical, or biological (enzymatic) conversion of woody biomass.

Productivity may also be improved by selecting genotypes best suited for growth and biomass accumulation in growing conditions found in marginal lands not suitable for higher value commodity crops. These breeding initiatives must address numerous biogeographic factors associated with genotype–environment interactions including physical, chemical, and biological components of soils such as nutrient levels and microbial diversity, as well as the overall climate and rainfall of the local geography.

While tree breeders are addressing biogeographic considerations relevant to growth, improvements in fiber quality and energy content are also major priorities for biomass feedstocks. Breeding objectives to improve woody feedstocks for heat generation can be quite different than for liquid fuels. For solid fuels for burning, the goal is to maximize energy content, while for liquid fuels the most important objective may be to reduce the recalcitrance of woody fiber to thermal, chemical, enzymatic, and biological breakdown.

4.2.1 Breeding for Cellulosic Ethanol

In recent years, emphasis has been placed on the development of SRWCs for the production of cellulosic ethanol. Woody tissues have a high density of polysaccharides, predominantly cellulose, that can be fermented into ethanol. While woody biomass has the potential to yield large amounts of ethanol, the general recalcitrance of woody tissue to digestion is the largest hurdle to realizing that potential.

Variations in cellulose, hemicellulose, and lignin content between species are well documented, but on average are approximately as 2:2:1, respectively. Among common cellulosic bioenergy feedstock plants, the amount of lignin in poplar, switchgrass, and corn stover are similar (approximately 18 %), while lignin in pine and wheat straw is higher (23–29 %). Interestingly though, holocellulose content in the woody biomass species pine and poplar (46–50 %) is greater than holocellulose content in maize, switch grass, and wheat straw (31–38 %) (Baker 1973).

In hardwood SRWCs the composition of hemicellulose varies greatly, typified by poplar which consists of 80 % xylans, 10 % mannans, with the rest accounted for by galactans and arabinogalactans (Mansfield 2009). Softwood SRWCs (such as pines) consist of approximately 50 % mannans, 30 % xylans, with the remaining 20 % made up of galactans and arabinogalactans (Harfouche et al. 2012). The composition of cell walls vary among species and varieties and represent a possible breeding target to attain the optimal composition for a desired energy use.

Genetic improvement through breeding can increase yields and reduce the overall cost of producing SRWCs biomass and support long-term sustainability of bioethanol production (Karp and Shield 2008). Economic modeling indicated that a 17 % increase in yield could increase the rate of return for SRWC biomass by 51 % (Buchholz and Volk 2011). Breeding of SRWCs for biofuels has been underway for over 30 years. Willow breeding as SRWC began in North America in the early 1980s at the University of Toronto (Zsuffa 1988). Various *Salix* species were tested and variations observed in plant biomass yields, moisture content and specific gravity traits associated among the species (Mosseler et al. 1988). Also investigated were pollination barriers, crossability relationships, and performance of willow hybrids (Mosseler and Papadopol 1989; Mosseler and Zsuffa 1989; Mosseler 1990). In the USA, willow breeding began with a large and diverse breeding collection of willows from across the Northeast and Midwestern USA (Kopp 2000; Kopp et al. 2001; Smart et al. 2005; Smart and Cameron 2008). Since 1998, more than 600 crosses were completed in the SUNY-ESF SRWC breeding program, with elite varieties being broadly tested in trials across North America (Smart and Cameron 2008; Smart and Cameron 2011). A gap remaining in our understanding of SRWC biomass production lies in the genotype–environment interactions that lead to significant changes in the performance of elite bioenergy varieties grown on different sites for willow (Kiernan et al. 2003). Genotype–environment interactions affecting biomass yields have been observed with both poplar and willow SRWC trials (Riemenschneider et al. 2001; Zalesny et al. 2009; Bergante et al. 2010) which could result from a wide range of biotic and/or abiotic environmental factors that limit growth potential, such as soil properties, precipitation, drainage, temperatures, pests and diseases, or weed competition.

There have also been numerous biotechnological strategies for the modification of woody tissue to reduce recalcitrance and to increase cellulosic ethanol yields, of which two have emerged as the most common. First, SRWC plant breeders have tried to make polysaccharides more amenable to breakdown (either by physical or microbial enzyme action, or a combination of both). A second strategy has been to reduce the amount or change the composition of lignin, which is the most recalcitrant compound inhibiting the release of fermentable sugars from woody biomass (Mansfield 2009).

4.2.2 *Modifying the Lignin Content of Woody Biomass*

Manipulating the lignin content in woody biomass has long been of interest to plant breeders and biotechnologists. Lignin interferes with the chemical breakdown of wood in the pulp and paper industry and reduces the digestibility of herbaceous forage plants by farm animals. Thus, researchers interested in the development of SRWCs for bioenergy have been motivated to reduce the effect of lignin on the breakdown of cellulosic materials. As a result, many of the genes coding for enzymes in the lignin pathway have been elucidated and have been down-regulated through genetic transformation in experiments with forest trees such as poplar and eucalyptus, as well as herbaceous plants such as tobacco and *Arabidopsis* (Anterola and Lewis 2002; Boerjan et al. 2003; Weng et al. 2008; Li et al. 2008; Barakat et al. 2011; Vanholme et al. 2012).

Two approaches to make SRWCs more efficient for biomass conversion to liquid fuels have been to reduce the overall amount of lignin in the wood or to modify the lignin content so that it is easier to chemically remove or to break down. Lignin monomer composition varies among species with gymnosperm lignin consisting mainly of G units, and angiosperm lignins containing both G and S units. Syringyl-(S-) lignin, which is higher in oxygen content than guaiacyl-(G-)lignin, is generally believed to be easier to be chemically removed (Mansfield 2009), although some have reported that lignin composition has little effect on the ease of cell-wall breakdown as opposed to lignin content (Wegrzyn et al. 2010). Many studies have been conducted on the modification of individual steps in the lignin biosynthetic pathway in the SRWC poplar. When genes early in the pathway have been manipulated in expression, differences in the composition of S- to G-lignin have been detected in cell walls. In the case of down-regulation of COMT or of ferulate 5-hydroxylase (F5H), the amount of S-lignin was greatly reduced, while over-expression of F5H led to increased S-lignin (Vanholme et al. 2012).

In *Populus*, pleiotropic effects have been recognized during the down-regulation of lignin pathway genes. These effects include the reduced removal of sugars from leaves and the subsequent reduction in the overall photosynthetic rate when the gene encoding *p*-coumaroyl shikimate 3′-hydroxylase was down-regulated (Coleman et al. 2008). A reduction in the amount of hemicellulose and an increase in vessel collapse and cavitation have been observed when phenylpropanoid acid pathway genes have been down-regulated (Leplé et al. 2007). The reduction of pleiotropic effects associated with lignin reduction in plants can probably be alleviated through genomic breeding. This may lead to optimized regulation of promoters and their target genes (Mizrachi et al. 2012) or by targeting pathway regulation hubs that can affect suites of genes involved in lignin biosynthesis through the phenylpropanoid pathway or pathways that provide precursor compounds to lignin biosynthesis (Mansfield 2009). An approach to increase lignin extractability in poplar without altering the lignin biosynthetic pathway has also been reported in which amino linkages are entered into the lignin polymer has shown promising results (Liang et al. 2008).

4.2.3 *Breeding for Maximization of Energy Content*

Although producing cellulosic ethanol has been an important goal of SRWC breeding, the use of woody crops as a combustible fuel is already a market for growers implemented in numerous markets across North America (Ellis 2012). When cellulose is burned it liberates a calorific value of approximately 17,000 kJ per kg of wood, while lignin liberates approximately 23,000 kJ per kg (Demirbas 2001). High lignin wood can be burned in co-gasification furnaces with coal and other fuels, or it can be directly burned on its own, for the production of heat, steam, or for electricity generation.

Perhaps the most direct strategy for increasing the thermal content of wood is to increase both the overall lignin content and increase the cellular density of the cell walls in wood (Demirbas 2001). While there may be limits to cell wall density, SRWCs can be bred to increase both cell wall density and to reduce the per cell lumen volume, which typically results in the thickening of the xylem. It appears that lignin synthesis in plant cell walls and cell wall thickening through increased xylem formation may be genetically associated. Numerous studies (Patzlaff et al. 2003; Goicoechea et al. 2005) have reported that several MYB transcription factors may be responsible for regulating lignin biosynthesis by binding to promoters for lignin biosynthesis genes. These same MYB transcription factors have been implicated in thickening of the xylem through increased deposition in the S2 layer of woody plant cell walls.

4.3 Genomics in Woody Biomass Improvement

Unlike annual agricultural crops with short generation times, traditional breeding methods for the genetic improvement of trees is challenged by long generation times, with one cycle often spanning a decade or more for forest trees (Harfouche et al. 2012).

The use of genomics has revolutionized biology and plant breeding has not been immune to these improvements. New technologies for genome sequencing and rapidly identifying genes in breeding populations allow researchers to capitalize on naturally occurring genetic variation (Fig. 4.2). To take full advantage of these technologies, SRWC breeders must be able to model associations of genotype-phenotype relationships. This relies on solid phenotypic data collected over several years for traits such as yield, fiber quality, and stress resistance. The ability to clonally propagate SRWC plants by cuttings is a great advantage, relative to forest trees, as it permits one genotype to be evaluated simultaneously over several different environments. On the other hand, however, the incidence of environmental stress is expected to be especially high for SRWC on marginal lands as a consequence of future climate predictions (Ellis 2012), which is challenging to predict and model.

4.3.1 Marker-Assisted Selection

Marker-Assisted Selection (MAS), the use of a genetic marker assigned to the region of a genome for the indirect selection of a breeding trait of interest, is a valuable resource for SRWC breeding (Grattapaglia and Resende 2011). Markers used in MAS may be morphological, biochemical, genetic, or based on many different types of genetic polymorphisms at the DNA level. Typically MAS uses genetic polymorphisms that have been previously established through quantitative trait loci (QTL) analysis. The major benefit of MAS is an overall decrease in the time required to complete selection steps in breeding. For trees MAS can abbreviate the selection

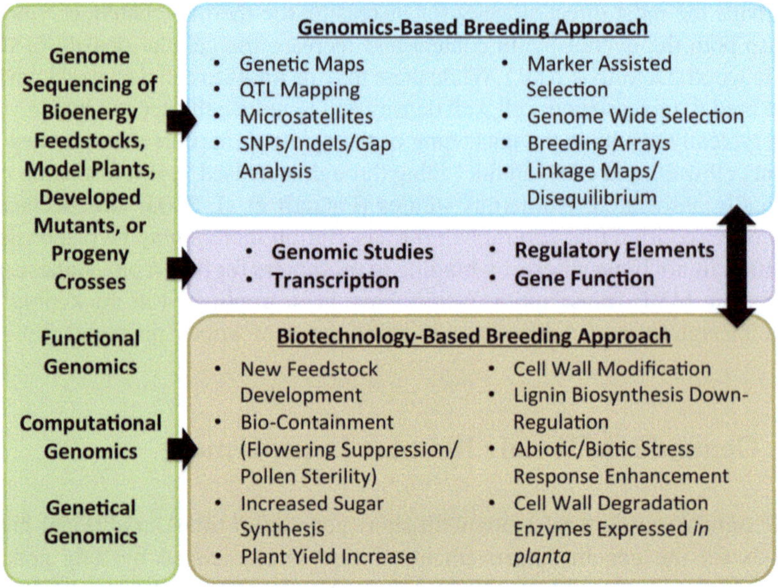

Fig. 4.2 Graphical representation of the use of genomics implemented in both genomic-based and biotechnology-based SRWC breeding approaches

cycle by many years, as MAS can be conducted as soon as leaves can be collected from seedlings, while phenotypic selection requires at least several years of evaluations of field trials. Some concerns regarding MAS have been that QTL analysis provides a small range of alleles that may be family-specific, and that there is weak linkage disequilibrium (LD) to work with in tree breeding populations (Resende et al. 2012). Forest trees have been notorious for having high heterozygosity (Neale and Ingvarsson 2008), weak LD (Lexer and Stölting 2012), and a large amount of rare alleles found at low frequencies (Eckert et al. 2009), which have made it challenging for breeders to identify phenotype-associated polymorphisms of large effect that are applicable beyond individual families (Quesada et al. 2010). Moving beyond QTL analysis, plant breeders and genomics groups are starting to identify genetic polymorphisms by association genetics or, when possible, with whole-genome sequence data (Neale and Kremer 2011).

4.3.2 The Future Is Here: Selection at the Whole Genome Level

Perhaps the most valuable asset to tree breeders from genomics will be tools for genome-wide association studies (GWAS) (Grattapaglia and Resende 2011). Unlike MAS, GWAS addresses large populations and takes both frequent and rare alleles into consideration (Resende et al. 2012). The overarching goal of incorporating GWAS into tree-improvement programs is to provide polymorphic information of broad value in selection. For GWAS to be valuable, sequence or marker data must be available for alleles across the genome and must represent population sizes large enough to determine association of traits with alleles at high levels of confidence, and preferably for common genetic factors that influence traits across families. Using GWAS, all allele variants in a breeding population can be determined, in theory. This is best accomplished for SRWC and forest trees having moderate genome sizes such as *Populus*, at 450 Mb, as opposed to large genomes such as conifers, at over 10 Gb, for which reference genomes are not yet completed (Resende et al. 2012).

The development and use of Genomic Selection (GS) tools that incorporate a predictive model using both genotypic and phenotypic data (Grattapaglia and Resende 2011) is an alternative to MAS. GS models are optimized to use all available genetic markers to predict the genomic landscape of future progeny. One model, the genomic breeding value, is a calculation developed using all the alleles at any given locus and estimating their breeding effect through linear regression (Resende et al. 2012). Optimal for SRWCs, GS is arguably the best method for traits expressed at different stages in the life cycle, with low heritability, or those not easily attainable through lower-throughput methods. The impact of GS in agronomic annual crop plants will be potentially smaller due to already existing advanced breeding strategies that can be implemented in months; it is expected that GS will have the greatest impact for long-lived plants with extended breeding cycles, typical of forest trees and the SRWCs developed from them (Resende et al. 2012). The impact of GS tools will remain strong as long as there are loci that have yet to be fully characterized (Grattapaglia and Resende 2011).

Recent breeding experiments using GS for forest trees have shown that predictive models can be developed, even with a low number of markers, for small effective breeding size populations that are comparable to traditional sized populations (Grattapaglia et al. 2009). With larger sizes of breeding populations and an increase in the LD within breeding populations, a higher marker density is required, which is provided by GS (Resende et al. 2012). The role of GS in breeding SRWCs is now being established, through gains in the reduction of selection times (Harfouche et al. 2012).

The largest impact of GS in the breeding of SRWCs may be in the reduction of breeding costs through the acceleration of superior phenotypes. It is typical for SRWC breeders to generate many crosses from numerous parents during multiyear experiments and GS provides ideal allele combinations that may reduce reliance on large breeding populations (Harfouche et al. 2012). A secondary benefit of the use of GS will be the reduction in breeding times associated with determining the best germplasm to use in coordinated crosses. Through both the reduction of breeding cost and reduced time to evaluate crosses, GS is poised to be the most widely used breeding strategy.

4.3.3 Next-Generation Sequencing in Tree Breeding

The development of next-generation sequencing (NGS) methods (Shendure and Ji 2008), resulting in greater throughput sequence production at lower cost, has drastically improved our ability to assess genomic variation for breeding efforts. Most importantly, NGS has made it possible to generate entire genome sequences for organisms much faster than previous sequencing technologies (Hill 2012). With NG technologies, allele variations can be directly identified in large breeding populations or, in the extreme case, even entire species (Tuskan et al. 2006). NGS techniques are used most frequently for two types of resequencing: for entire genomes of individuals in a species that has a reference genome, such as a *Populus* breeding population (Cohen et al. 2010), or to resequence a suite of genes or regions of interest from large breeding populations to detect variants. There are numerous types of variants that may be identified by resequencing, including single nucleotide polymorphisms (SNPs) or insertion and deletion events (indels), to more extensive variants such as transposable elements (Harfouche et al. 2012) or rearrangements and structural variants (SVs) (Villar et al. 2011). Trait variation, as identified by GWAS, is best suited to QTL or individual candidate genes and common alleles, but can typically only identify a small fraction of the actual variation in individuals. The use of resequencing approaches has the added benefit in that rare functional variants can also be identified (Elshire et al. 2011).

NGS technologies are expected to change rapidly in the future such that they will become routine in breeding applications with forest trees. This has already becoming the case in *Populus*, where microarrays and SNP genotyping chips have been replaced by methods for genotyping-by-sequencing (GBS; Elshire et al. 2011; Davey et al. 2011). New resequencing technologies are most applicable in species with sequenced genomes, but one advantage of GBS technologies is that the initial investment of genome sequencing or SNP chip development are not mandatory prerequisites for genotyping in forest trees. Forest trees are especially prone to weak LD and the identification of a large number of low cost markers from GBS are a solution to a lack of genetic markers (Elshire et al. 2011).

4.3.4 Integration of Molecular Data in Breeding Technologies

Information derived from new sequencing technologies in the form of genomic and transcriptomic data will progressively be integrated with new technologies in the areas of proteomics and metabolomics in SRWC breeding initiatives (Cohen et al. 2010; Hamanishi et al. 2010). Integrating data collected from breeding populations developed by GS or allele variants from GWAS studies, holistic information from the integration of many forms of information from the genetic to cell chemistry will increase productivity and quality of cultivars developed by SRWC plant breeders. The fields of proteomics and metabolomics are still in the developmental stage for forest trees (Nilsson et al. 2010), but there is much promise for SRWC breeding initiatives.

Transcriptomics, which include methods such as microarrays and RNA-Seq, are comprehensive ways to measure genome-wide expression coverage and change (Baginsky et al. 2010; Qiu et al. 2011). RNA-Seq in particular is contributing to the understanding of transcript profiling at more highly resolved levels than microarrays (Wang and Brummer 2012), through the increasing availability of NGS methods. Additionally, transcriptomics can be used to understand both genetic and epigenetic regulation of gene expression in forest trees and how yield may be affected by specific environments. Forest breeders can implement transcriptional markers to predict heterosis in mating pairs as has been accomplished in *Arabidopsis* (Lister et al. 2008).

Proteomics is a key technological development in understanding the role of proteins in the accumulation of biomass or environmental stress response. Proteomic methods can yield information regarding structural conformation and variation from posttranslational modification or quantitative traits of proteins such as stability and tissue-specific protein accumulation (Gstaiger and Aebersold 2009). Proteomics benefits SRWC breeding because it can transcend measurements of mRNA levels to predictions of gene products and posttranslational modification events (Baginsky et al. 2010).

Metabolic markers are in some ways the ultimate screening tools, as they have been shown to be more highly correlated with phenotype than transcript levels (Boerjan 2005). As a result, monitoring metabolites may be a more rapid way of advancing phenotypic selection for SRWC breeders than following gene expression alone. Metabolic markers have been developed in *Arabidopsis* where correlations between the metabolome and biomass accumulation were observed (Meyer et al. 2012). The fact that metabolite levels are sensitive to environmental variation has been both an advantage and caveat with metabolic markers, however, relative to the stability of genetic markers.

Integrating transcriptomics, proteomics, and metabolomics into SRWC breeding programs will complete the already vibrant functional genomics toolbox for the development of superior performing cultivars. This will enable a systems biology approach in plant improvement that will be crucial to the understanding of regulatory mechanisms for complex traits such as environmental stress resistance and characterization of hormone pathways (Neale and Ingvarsson 2008). Integration of these genomic tools will enhance our knowledge of the complexity of the genes, proteins, and metabolites involved in cellular networks governing phenotypic traits (Baginsky et al. 2010). The end result of the integration of transcriptomics, metabolomics, and proteomics with GBS and GWAS is the goal of exploiting natural biological and genetic variation to expand the tree improvement toolbox available to SRWC breeders.

4.4 Biotechnology in SRWC Improvement: The Use of Genetic Engineering

4.4.1 What Is Genetic Engineering

Genetic engineering (GE) can also be used to complement conventional breeding programs for SRWC species. Incorporating GE with existing traditional and genomics-based breeding programs allows breeders to take advantage of natural genetic variation and introduce genes for traits which may not be available within the species, or be difficult to introduce via other breeding methods (Busov et al. 2005). The important SRWC target of high yields of digestible woody biomass is increasingly feasible using GE methods (Grattapaglia et al. 2009; Mizrachi et al. 2012).

Numerous strategies can be implemented using GE. One approach involves directly adding or modifying genes responsible for increased plant growth, improved stress tolerance or environmental adaptability, or increased secondary metabolites or byproducts (Grattapaglia et al. 2009). Another approach can be to express transcription factors to up-regulate expression of target genes to accomplish a specific trait or suite of traits (Mizrachi et al. 2012). The combination of these strategies can be used in tandem with many different genes, in a process known as trait stacking.

Genetic engineering can support plant breeding in numerous ways. It is particularly useful in woody plants that have long juvenile periods before breeding can be conducted (Harfouche et al. 2011). Unwanted phenotypic variation is often observed in breeding populations resulting from introgression, segregation and recombination, and single-gene traits have been known to break down in elite breeding genotypes after years of cultivation. These issues could be addressed through GE by the use of dominant and highly expressed genes, as well as multigenic approaches (Harfouche et al. 2011). The limiting factor for GE with trees has been the understanding of the genetic basis of complex traits. However, complex traits, such as disease resistance and nutrient use efficiency, which have challenged traditional breeding of SRWCs, may soon be approachable with multigenic GE techniques.

4.4.2 Implementation of Genetic Engineering in SRWC Breeding

Genetic engineering will feature prominently when improvement of SRWCs is needed for traits that are not observed in a breeding population or the species as a whole, or when a lack of natural variation in a trait makes it particularly difficult to use GS for selection. One example of when GE could be most useful is the addition of a single gene—for which there is no analogue in the breeding plant's respective genome—such as genes to confer resistance to an exotic pest or disease that could take far too long to incorporate from related species by backcrossing (Quesada et al. 2010). If there is no genetic variation for a trait of interest, GE may be used to introduce sources of variation from other species. Lastly, GE methods may be able to produce early flowering in some SRWC plants allowing for faster breeding cycles (Flachowsky et al. 2009). In contrast, it is more likely that GE methods will be used to reduce, delay, or completely knock out flowering in cases where containment of gene flow from GE plants, or from large-scale plantations of exotic species or hybrids, to other tree populations would be a concern.

4.4.2.1 Cell Wall Modification Through Genetic Engineering Means

Perhaps the most exciting use of GE for SRWCs will be the modification of cell wall composition. Modifying the components of the cell wall may increase the amount of energy attained from SRWC material or may reduce the amount of recalcitrance associated with conversion of plant tissues into energy or biomaterials (Campbell et al. 2003). Studies incorporating GE have influenced cell wall development by modification of a β-1,4-endoglucanase (*CEL1*) responsible for enhanced growth in hybrid *Populus* (Shani et al. 2006). Additional studies have shown that using GE to add and express a cytosolic glutamine synthetase (*GS1*) in hybrid *Populus* derived from *Pinus* increased cell wall biomass accumulation, most notably in leaf tissue (Bernard and Habash 2009). The expression of the transgenic *GS1* was verified with increased enzyme activity along with an overall increase in amino acid concentration (Man et al. 2005) and drought tolerance (El-Khatib et al. 2004). Numerous studies have used GE methods to modify the content, structure and composition (G versus S subunits) of lignin in the cell walls of SRWC plants (Fu et al. 2003; Boerjan 2005). Novel approaches such as introducing amino-linkages within the lignin polymer by targeting of TYR-rich proteins to the cell wall of developing xylem in poplar (Liang et al. 2008) also show promise.

4.4.2.2 Stress Resistance Traits Enhanced Through Genetic Engineering Means

Genes involved in leaf size and structure, stem development, timing of bud flush, and leaf senescence all may influence biomass accumulation, rotation time, and growth rate. It is also possible that genes that confer stress tolerance or delay senescence could also improve growth under environmental stress conditions or extend the growing season, and it has been shown that suppression of a gene, deoxyhypusine synthase (DHS) which is part of the stress-response pathway, increases vegetative and reproductive growth in the model plant *Arabidopsis* (Duguay et al. 2007).

For example, transgenics has led to the development of cold-tolerant hybrid eucalyptus trees (*Eucalyptus grandis* X *Eucalyptus urophylla*) that are awaiting field trials (Hinchee et al. 2009). Introduced genes being tested in *Populus* include the *Populus tremula* and *Arabidopsis* stable protein 1 (SP-1) gene, as well as genes involved in metabolic processes responsive to drought, redox proteins, transporter proteins, signal transduction proteins, and transcription factors (Polle et al. 2006).

Another area for improvement is the engineering of belowground traits. Roots are the key to a second "Green Revolution" in agriculture (Evenson and Gollin 2003), and the same holds true for establishing SRWC trees with higher productivity. Although efforts to boost root growth by conventional breeding are underway in some species, GE may prove to be more efficient. For example, discovery of the WUSCHEL-related homeobox gene (WOX11), which is involved in the activation of crown root emergence and growth (van der Graaff et al. 2009), and UPBEAT1, which is a key transcription factor that controls the transition between cell proliferation and differentiation in Arabidopsis roots (Yokawa et al. 2013), suggests that it may be possible to design root systems tailored to certain environments. Thus, it may be possible to reduce the time needed for root systems to become established on recalcitrant forest tree species, therefore reducing the time to harvest. By simultaneously engineering several genes involved in root growth, GE can support research to obtain a more rapid understanding of complex, multigenic traits, such as adventitious rooting.

4.4.2.3 Development of Mutant Lines Through Genetic Engineering

As a complement to transgenic, forward-genetic approaches such as cisgenics and intragenics, implemented as activation tagging, are used to develop mutant lines and identify genes controlling specific traits in SRWCs (Ahuja 2011). Biotechnology programs have the goal of developing thousands of activation-tagged mutant lines for *Populus* in this fashion.

Activation-tagged genes have potential use for the further development of transgenic and cisgenic GE SRWC plants. Additionally, activation tagging may elucidate novel targets for improved SRWC breeding. The majority of GE work on activation tagging has been conducted in maize, Arabidopsis, and more recently poplar, with numerous genes showing promise for breeding using reverse-genetic approaches. Most notably, the gene LATERAL ORGAN BOUNDARIES DOMAIN (*LBD*) showed positive regulation of phloem formation during secondary growth (Yordanov et al. 2010). In another study, a gene responsible for coding gibberellin catabolism (GA 2-oxidase) was found to regulate tree height (Busov et al. 2003).

Another approach to identifying phenotypes important for SRWC improvement is through the use of transient transformation by Virus Induced Gene Silencing (VIGS). The great potential of VIGS as a powerful functional genomics platform for discovery of genes effecting plant cell wall formation was demonstrated in tobacco by (Zhu et al. 2010). Efforts to establish a VIGS functional genomics platform in poplar for SRWC improvement are also underway (S. Joshi, personal communication).

Another method for the development of mutant lines that has potential use with SRWCs is the implementation of Zinc finger nucleases (ZFNs). This strategy uses ZFN enzymes to target a specific locus by creating a double-stranded DNA break that can include both targeted insertions and point mutations (Durai et al. 2005). The major benefit of using ZFNs is avoiding the randomness found with traditional transgene insertion methods. The use of ZFNs has yet to be fully realized in SRWCs, but the potential is there for the enhancement of breeding and genomic selection.

4.4.2.4 The Use of Small RNA Interference for Gene Silencing

Perhaps the most widely used method for targeted gene silencing is the use of small non-transcribed RNA fragments identified as RNA interference (RNAi). These methods have been implemented with both direct introduction of designed RNA fragments or microRNAs (miRNAs); both methods take advantage of a plant's innate small interfering RNA (siRNA) processing pathway (Li et al. 2011). The most notable use of RNAi with SRWCs has been the suppression of genes in *Populus* including the STYLISH1 (*STY1*) and SHORT INTER-NODES (*SHI*) genes, which showed increased shoot and root growth and increased the amount of xylem tissue (Zawaski et al. 2011). The insertion of miR156 class miRNA has been effective in down-regulating the SQUAMOSA PROMOTER BINDING PROTEIN (*SBP*) modulating the growth of lateral branches and meristems (Chuck et al. 2007). An in-depth analysis of miRNA genes and gene families in poplar (Barakat et al. 2007), provided a gene silencing tool set for poplar for SRWC development. This study showed that miRNA family sizes were generally expanded in *Populus trichocarpa* relative to rice and Arabidopsis, with 48 new miRNA families that could be *Populus-specific*, and almost half of the genes targeted by non-conserved miRNAs appearing to be *Populus*-specific. The putative targets of non-conserved miRNAs in poplar included several involved in development, resistance to stress, and other cellular processes. Further studies with *Populus euphratica* have yielded a suite of miRNAs associated with long-term drought stress that could be implemented in plant breeding initiatives (Li et al. 2011).

4.4.3 Regulatory Requirements and Associated Risk Assessment for Biotech Trees

The use of GE strategies may prove to be highly valuable in the development and implementation of biofuels from SRWCs and may provide environmental, social, and economic benefits as well (Ahuja 2011). The use of GE in SRWC breeding will allow novel traits to be introduced in a timely fashion not capable with traditional breeding techniques.

Numerous concerns must be addressed before GE modified SRWCs may be implemented in practice, however. Perhaps the most important concern is the dispersal of GE pollen, seeds, or other vegetative propagules (Strauss et al. 2004). SRWCs are generally wind pollinated and it is possible that GE pollen can travel long distances and pollinate receptive native species. Although traits being bred into bioenergy SRWCs—such as improved yield, wood composition, or abiotic and biotic stress resistance—can be considered low risk, the overall goal or requirement for large-scale crop release could be to eliminate all possibility of escape of GE modified genes. Many SRWCs are clonally propagated and do not reach flowering before they are harvested, but as a precaution, GE plants of SRWCs may be produced in genetic backgrounds that are pollen sterile, such a triploids (Busov et al. 2005). This can be implemented through the addition of knockout genes for sterility which prevent the formation of pollen without affecting the general vigor of the plant (Strauss et al. 2004; Ahuja 2011).

The use of GE in breeding of SRWCs for bioenergy purposes will require regulatory guidance and approval prior to commercialization. While field tests of GE modified trees have been implemented in numerous countries, the USA has no GE trees in field production (Harfouche et al. 2012). The only GE SRWC trees grown industrially are insect resistant poplar genotypes in China (Ahuja 2011), while freeze-tolerant Eucalyptus lines await approvals (Harfouche et al. 2011), and currently the only non-regulated GE trees in the USA are papaya and plum.

Regulations for SRWC plants bred using GE technology should meet clearly defined simple standards agreed upon by global contributors. Studies considering the safety and implementation of GE plants, as well as a thorough cost–benefit analysis should be a priority for SRWCs (Strauss et al. 2004). There is a growing consensus among the tree improvement community that guidelines for all plants developed by genomic or biotechnological means should address traits and not the method by which superbly performing cultivars were developed (Harfouche et al. 2011).

4.5 Conclusion

To meet future renewable energy standards, clearly designated, sustainable plant-based bioenergy production benchmarks consisting of a variety of bioenergy sources are needed, of which forest trees and SRWCs must play a significant role. Poised to make minor contributions at current levels of production, trees will have to be bred for higher productivity and greater resistance to abiotic and biotic stresses associated with marginal lands in order to substantially contribute towards future energy benchmarks.

The uses of SRWCs and forest trees in bioenergy are diverse as they may provide energy in a wide range of uses such as cellulosic ethanol as a liquid fuel or through combustion to generate electric power. Additionally, unlike other bioenergy plants, SRWCs may be used as raw materials in the pulp and paper industry, as well as building materials, and in carbon sequestration for climate change abatement. Purpose-bred SRWCs grown in high-rotation plantations will alleviate pressure to harvest native forests. In order to meet all these demands, SRWCs are being improved to optimally meet these criteria.

The integration of genomics, biotechnology, and SRWC breeding has great potential to improve the yield and sustainability of bioenergy production. Dedicated SRWC breeding initiatives can make large contributions to climate change abatement and economies, both at local and global scales. For the ultimate economic and environmental impacts to be achieved, the development of SRWC breeding must be supported by government and private funding, meet regulatory acceptance, and be supported by the public at large.

References

Ahuja MR (2011) Fate of transgenes in the forest tree genome. Tree Genet Genomes 7:221–230

Anterola AM, Lewis NG (2002) Trends in lignin modification: a comprehensive analysis of the effects of genetic manipulations/mutations on lignification and vascular integrity. Phytochemistry 61:221–294

Baginsky S, Hennig L, Zimmermann P, Gruissem W (2010) Gene expression analysis, proteomics, and network discovery. Plant Physiol 152:402–410

Baker AJ (1973) Effect of lignin on the in vitro digestibility of wood pulp. J Anim Sci 35: 768–771

Barakat A, Wall PK, DiLoreto S, dePamphilis CW, Carlson JE (2007) Conservation and divergence of microRNAs in Populus. BMC Genomics 8:481

Barakat A, Yassin NBM, Park JS, Choi A, Herr JR, Carlson JE (2011) Comparative and phylogenomic analyses of cinnamoyl-CoA reductase and cinnamoyl-CoA-reductase-like gene family in land plants. Plant Sci 181:249–257

Bergante S, Facciotto G, Minotta G (2010) Identification of the main site factors and management intensity affecting the establishment of short-rotation-coppices (SRC) in northern Italy through stepwise regression analysis. Cent Eur J Biol 5:522–30

Bernard SM, Habash DZ (2009) The importance of cytosolic glutamine synthetase in nitrogen assimilation and recycling. New Phytol 182:608–620

Boerjan W (2005) Biotechnology and the domestication of forest trees. Curr Opin Biotechnol 16:159–166

Boerjan W, Ralph J, Baucher M (2003) Lignin biosynthesis. Annu Rev Plant Biol 54:519–546

Buchholz T, Volk TA (2011) Improving the profitability of willow crops—Identifying opportunities with a crop budget model. Bioenerg Res 4:85–95

Busov VB, Brunner AM, Meilan R, Filichkin S, Ganio L, Gandhi S, Strauss SH (2005) Genetic transformation: a powerful tool for dissection of adaptive traits in trees. New Phytol 167:9–18

Busov VB, Meilan R, Pearce DW, Ma C, Rood SB, Strauss SH (2003) Activation tagging of a dominant gibberellin catabolism gene (GA 2-oxidase) from poplar that regulates tree stature. Plant Physiol 132:1283–1291

Campbell MM, Brunner AM, Jones HM, Strauss SH (2003) Forestry's fertile crescent: the application of biotechnology to forest trees. Plant Biotechnol J 1:141–154

Chuck G, Cigan AM, Saeteurn K, Hake S (2007) The heterochronic maize mutant Corngrass1 results from overexpression of a tandem microRNA. Nat Genet 39:544–549

Cobb WR, Will RE, Daniels RF, Jacobson MA (2008) Aboveground biomass and nitrogen in four short-rotation woody crop species growing with different water and nutrient availabilities. For Ecol Manag 255:4032–4039

Cohen D, Bogeat-Triboulot M-B, Tisserant E, Balzergue S, Martin-Magniette M-L, Lelandais G, Ningre N, Renou J-P, Tamby J-P, Le Thiec D et al (2010) Comparative transcriptomics of drought responses in Populus: a meta-analysis of genome-wide expression profiling in mature leaves and root apices across two genotypes. BMC Genomics 11:630

Coleman HD, Samuels AL, Guy RD, Mansfield SD (2008) Perturbed lignification impacts tree growth in hybrid poplar–a function of sink strength, vascular integrity, and photosynthetic assimilation. Plant Physiol 148:1229–1237

Davey JW, Hohenlohe PA, Etter PD, Boone JQ, Catchen JM, Blaxter ML (2011) Genome-wide genetic marker discovery and genotyping using next-generation sequencing. Nat Publ Group 12:499–510

Demirbas A (2001) Relationships between lignin contents and heating values of biomass. Energ Convers Manage 42:183–188

Duguay J, Jamal S, Liu Z, Wang T-W, Thompson JE (2007) Leaf-specific suppression of deoxyhypusine synthase in Arabidopsis thaliana enhances growth without negative pleiotropic effects. Journal of Plant Physiology 164(4) 408–420

Durai S, Mani M, Kandavelou K, Wu J, Porteus MH, Chandrasegaran S (2005) Zinc finger nucleases: custom-designed molecular scissors for genome engineering of plant and mammalian cells. Nucleic Acids Res 33:5978–5990

Eckert AJ, Bower AD, Wegrzyn JL, Pande B, Jermstad KD, Krutovsky KV, St Clair JB, Neale DB (2009) Association genetics of coastal Douglas fir (*Pseudotsuga menziesii* var. *menziesii*, Pinaceae). I. Cold-hardiness related traits. Genetics 182:1289–1302

El-Khatib RT, Hamerlynck EP, Gallardo F, Kirby EG (2004) Transgenic poplar characterized by ectopic expression of a pine cytosolic glutamine synthetase gene exhibits enhanced tolerance to water stress. Tree Physiol 24:729–736

Ellis BE (2012) Bringing trees into the fuel line. New Phytol 194:1–3

Elshire RJ, Glaubitz JC, Sun Q, Poland JA, Kawamoto K, Buckler ES, Mitchell SE (2011) A robust, simple genotyping-by-sequencing (GBS) approach for high diversity species. PLoS One 6:e19379

Evenson RE, Gollin D (2003) Assessing the impact of the green revolution 1960 to 2000. Science 300(5620):758–762

Flachowsky H, Hanke MV, Peil A, Strauss SH, Fladung M (2009) A review on transgenic approaches to accelerate breeding of woody plants. Plant Breed 128:217–226

Fu J, Sampalo R, Gallardo F, Cánovas FM, Kirby EG (2003) Assembly of a cytosolic pine glutamine synthetase holoenzyme in leaves of transgenic poplar leads to enhanced vegetative growth in young plants. Plant Cell Environ 26:411–418

Goicoechea MM, Lacombe EE, Legay SS, Mihaljevic SS, Rech PP, Jauneau AA, Lapierre CC, Pollet BB, Verhaegen DD, Chaubet-Gigot NN et al (2005) EgMYB2, a new transcriptional activator from Eucalyptus xylem, regulates secondary cell wall formation and lignin biosynthesis. Plant J 43:553–567

Grattapaglia D, Resende MDV (2011) Genomic selection in forest tree breeding. Tree Genet Genomes 7:241–255

Grattapaglia D, Plomion C, Kirst M, Sederoff RR (2009) Genomics of growth traits in forest trees. Curr Opin Plant Biol 12:148–156

Gstaiger M, Aebersold R (2009) Applying mass spectrometry-based proteomics to genetics, genomics and network biology. Nat Rev Genet 10:617–627

Hamanishi ET, Raj S, Wilkins O, Thomas BR, Mansfield SD, Plant AL, Campbell MM (2010) Intraspecific variation in the Populus balsamifera drought transcriptome. Plant Cell Environ 33:1742–1755

Harfouche A, Meilan R, Altman A (2011) Tree genetic engineering and applications to sustainable forestry and biomass production. Trends Biotechnol 29:9–17

Harfouche A, Meilan R, Kirst M, Morgante M, Boerjan W, Sabatti M, Scarascia MG (2012) Accelerating the domestication of forest trees in a changing world. Trends Plant Sci 17:64–72

Herr JR (2011) Bioenergy from trees. New Phytol 192:313–315

Hill WG (2012) Quantitative genetics in the genomics era. Curr Genomics 13:196–206

Hinchee M, Rottmann W, Mullinax L, Zhang C, Chang S, Cunningham M, Pearson L, Nehra N (2009) Short-rotation woody crops for bioenergy and biofuels applications. In Vitro Cell Dev Biol Plant 45:619–629

Karp A, Shield I (2008) Bioenergy from plants and the sustainable yield challenge. New Phytol 179:15–32

Kiernan BD, Volk TA, Tharakan PJ, Nowak CA, Phillipon SP, Abrahamson LP, White EH (2003) Clone-site testing and selections for scale-up plantings. SUNY College of Environmental Science and Forestry, Syracuse, NY. http://www.esf.edu/willow/reports.htm. Accessed 24 Jul 2013

Kopp RF (2000) Genetic improvement of Salix using traditional breeding and AFLP fingerprinting. Thesis, SUNY College of Environmental Science and Forestry, Syracuse, NY. pp 175

Kopp RF, Smart LB, Maynard CA, Isebrands JG, Tuskan GA, Abrahamson LP (2001) The development of improved willow clones for eastern North America. Forest Chron 77:287–92

Leplé J-C, Dauwe R, Morreel K, Storme V, Lapierre C, Pollet B, Naumann A, Kang K-Y, Kim H, Ruel K et al (2007) Downregulation of cinnamoyl-coenzyme A reductase in poplar: multiple-

level phenotyping reveals effects on cell wall polymer metabolism and structure. Plant Cell 19:3669–3691
Lexer C, Stölting KN (2012) Whole genome sequencing (WGS) meets biogeography and shows that genomic selection in forest trees is feasible. New Phytol 196:652–654
Li B, Qin Y, Duan H, Yin W, Xia X (2011) Genome-wide characterization of new and drought stress responsive microRNAs in Populus euphratica. J Exp Bot 62:3765–3779
Li X, Weng J-K, Chapple C (2008) Improvement of biomass through lignin modification. Plant J 54:569–581
Liang H, Frost CJ, Wei X, Brown NR, Carlson JE, Tien M (2008) A novel approach toward lignin modification to facilitate cellulosic ethanol production: introducing a tyrosine-rich cell wall peptide gene in poplar. Clean 36(8):662–668
Lister R, O'Malley RC, Tonti-Filippini J, Gregory BD, Berry CC, Millar AH, Ecker JR (2008) Highly integrated single-base resolution maps of the epigenome in Arabidopsis. Cell 133:14–14
Man H-M, Boriel R, El-Khatib R, Kirby EG (2005) Characterization of transgenic poplar with ectopic expression of pine cytosolic glutamine synthetase under conditions of varying nitrogen availability. New Phytol 167:31–39
Mansfield SD (2009) Solutions for dissolution–engineering cell walls for deconstruction. Curr Opin Biotechnol 20:286–294
Mead DJ (2005) Forests for energy and the role of planted trees. Crit Rev Plant Sci 24:407–421
Meyer RS, DuVal AE, Jensen HR (2012) Patterns and processes in crop domestication: an historical review and quantitative analysis of 203 global food crops. New Phytol 196:29–48
Mizrachi E, Mansfield SD, Myburg AA (2012) Cellulose factories: advancing bioenergy production from forest trees. New Phytol 194:54–62
Mosseler A, Zsuffa L, Stoehr MU, Kenney WA (1988) Variation in biomass production, moisture content, and specific gravity in some North American willows (*Salix* L.). Can J For Res 18:1535–40
Mosseler A, Papadopol CS (1989) Seasonal isolation as a reproductive barrier among sympatric *Salix* species. Can J Bot 67:2563–70
Mosseler A, Zsuffa L (1989) Sex expression and sex ratios in intra- and interspecific hybrid families of *Salix* L. Silvae Genet 38:12–17
Mosseler A (1990) Hybrid performance and species crossability relationships in willows (*Salix*). Can J Bot 68:2329–38
Neale DB (2007) Genomics to tree breeding and forest health. Curr Opin Genet Dev 17:539–544
Neale DB, Ingvarsson PK (2008) Population, quantitative and comparative genomics of adaptation in forest trees. Curr Opin Plant Biol 11:149–155
Neale DB, Kremer A (2011) Forest tree genomics: growing resources and applications. Nat Publ Group 12:111–122
Nilsson R, Bernfur K, Gustavsson N, Bygdell J, Wingsle G, Larsson C (2010) Proteomics of plasma membranes from poplar trees reveals tissue distribution of transporters, receptors, and proteins in cell wall formation. Mol Cell Proteomics 9:368–387
Patzlaff AA, McInnis SS, Courtenay AA, Surman CC, Newman LJL, Smith CC, Bevan MWM, Mansfield SS, Whetten RWR, Sederoff RRR et al (2003) Characterisation of a pine MYB that regulates lignification. Plant J 36:743–754
Polle A, Altman A, Jiang X (2006) Towards Genetic Engineering for drought tolerance in trees. Pages 275–297. From "Tree Transgenesis" Edited by M Fladung & D Ewald, Springer (Berlin, Heidelberg)
Qiu QQ, Ma TT, Hu QQ, Liu BB, Wu YY, Zhou HH, Wang QQ, Wang JJ, Liu JJ (2011) Genome-scale transcriptome analysis of the desert poplar, *Populus euphratica*. Tree Physiol 31: 452–461
Quesada T, Gopal V, Cumbie WP, Eckert AJ, Wegrzyn JL, Neale DB, Goldfarb B, Huber DA, Casella G, Davis JM (2010) Association mapping of quantitative disease resistance in a natural population of loblolly pine (*Pinus taeda* L.). Genetics 186:677–686

Resende MFR, Muñoz P, Acosta JJ, Peter GF, Davis JM, Grattapaglia D, Resende MDV, Kirst M (2012) Accelerating the domestication of trees using genomic selection: accuracy of prediction models across ages and environments. New Phytol 193:617–624

Riemenschneider DE, Berguson WE, Dickmann DI, Hall RB, Isebrands JG, Mohn CA, Stanosz GR, Tuskan GA (2001) Poplar breeding and testing strategies in the north-central US: demonstration of potential yield and consideration of future research needs. Forest Chron 77:245–53

Shani ZZ, Dekel MM, Roiz LL, Horowitz MM, Kolosovski NN, Lapidot SS, Alkan SS, Koltai HH, Tsabary GG, Goren RR et al (2006) Expression of endo-1,4-beta-glucanase (cel1) in *Arabidopsis thaliana* is associated with plant growth, xylem development and cell wall thickening. Plant Cell Rep 25:1067–1074

Shendure J, Ji H (2008) Next-generation DNA sequencing. Nat Biotech 26:1135–1145

Simmons BA, Loque D, Blanch HW (2008) Next-generation biomass feedstocks for biofuel production. Genome Biol 9:242

Sims REH, Venturi P (2004) All-year-round harvesting of short rotation coppice *Eucalyptus* compared with the delivered costs of biomass from more conventional short season, harvesting systems. Biomass Bioenergy 26:27–37

Smart LB, Cameron KD (2008) Genetic improvement of willow (*Salix* spp.) as a dedicated bioenergy crop. In: Vermerris WE (ed) Genetic improvement of bioenergy crops. Springer Science, New York, NY, pp 347–76

Smart LB, Cameron KD (2011) Shrub willow (*Salix* spp.) bioenergy crops. In: Kole C, Joshi S, Shonnard D (eds) Handbook of bioenergy crop plants. Taylor and Francis, Boca Raton, FL, In press

Smart LB, Volk TA, Lin J, Kopp RF, Phillips IS, Cameron KD, White EH, Abrahamson LP (2005) Genetic improvement of shrub willow (*Salix* spp.) crops for bioenergy and environmental applications in the United States. Unasylva 56:51–55

Strauss SH, Brunner AM, Busov VB, Ma C, Meilan R (2004) Ten lessons from 15 years of transgenic Populus research. Forestry 77:455–465

Tuskan GA, DiFazio S, Jansson S, Bohlmann J, Grigoriev I, Hellsten U, Putnam N, Ralph S, Rombauts S, Salamov A et al (2006) The genome of black cottonwood, Populus trichocarpa (Torr. & Gray). Science 313:1596–1604

van der Graaff E, Laux T, Rensing SA (2009) The WUS homeobox-containing (WOX) protein family. Genome Biol 10:248

Vanholme R, Morreel K, Darrah C, Oyarce P, Grabber JH, Ralph J, Boerjan W (2012) Metabolic engineering of novel lignin in biomass crops. New Phytol 196:978–1000

Villar E, Klopp C, Noirot C, Novaes E, Kirst M, Plomion C, Gion J-M (2011) RNA-Seq reveals genotype-specific molecular responses to water deficit in *Eucalyptus*. BMC Genomics 12:538–538

Wang Z-Y, Brummer EC (2012) Is genetic engineering ever going to take off in forage, turf and bioenergy crop breeding? Ann Bot 110:1317–1325

Wegrzyn JL, Eckert AJ, Choi M, Lee JM, Stanton BJ, Sykes R, Davis MF, Tsai C-J, Neale DB (2010) Association genetics of traits controlling lignin and cellulose biosynthesis in black cottonwood (Populus trichocarpa, Salicaceae) secondary xylem. New Phytol 188:515–532

Weng J-K, Li X, Bonawitz ND, Chapple C (2008) Emerging strategies of lignin engineering and degradation for cellulosic biofuel production. Curr Opin Biotechnol 19:7–7

Yokawa K, Kagenishi T, Baluška F (2013) Root photomorphogenesis in laboratory-maintained *Arabidopsis* seedlings. Trends Plant Sci 18(3):117–119

Yordanov YSY, Regan SS, Busov VV (2010) Members of the lateral organ boundaries domain transcription factor family are involved in the regulation of secondary growth in Populus. Plant Cell 22:3662–3677

Zalesny RS, Hall RB, Zalesny JA, McMahon BG, Berguson WE, Stanosz GR (2009) Biomass and genotype x environment interactions of *Populus* energy crops in the Midwestern United States. Bioenerg Res 2:106–22

Zawaski C, Kadmiel M, Ma C, Gai Y, Jiang X, Strauss SH, Busov VB (2011) Short internodes-like genes regulate shoot growth and xylem proliferation in Populus. New Phytol 191:678–691

Zhu X, Pattathil S, Mazumder K, Brehm A, Hahn MG, Dinesh-Kumar SP, Joshi CP (2010) Virus-induced gene Silencing offers a functional genomics platform for studying plant cell wall formation. Molecular Plant 3(5):818–833

Zsuffa L (1988) A review of the progress in selecting and breeding North American Salix species for energy plantations at the Faculty of Forestry, University of Toronto, Canada. International Energy Agency willow breeding symposium, Uppsala, Sweden, 41–51

Chapter 5
Biomass Harvesting Systems and Analysis

Jingxin Wang, Damon Hartley, and Weiguo Liu

5.1 Harvest System Overview

Harvesting of trees or woody biomass, for the production of lumber, for production of paper, or as a feedstock for bioenergy, is accomplished through the integrated activities of a number of people and machines. The integration of equipment designed to perform disparate tasks forms what is referred to as a harvest system. In the harvest system, each task is performed in concert with the goals of the subsequent tasks, the overall objective being to prepare the raw material for transport to the processing facility.

Timber harvesting systems exist in many forms and configurations (Stenzel et al. 1985; Stokes and Schilling 1997). Each system, while differing in machinery used and products produced, completes the five basic harvesting phases of cutting, primary transport, processing, loading, and secondary transport. Each harvesting phase is made up of one or more steps or sequences of activities that are called "harvest functions."

Harvesting systems begin with felling or severing trees or woody biomass from the stump which is followed by primary transport, processing, loading, and ends with secondary transport. Processing can happen either before or after primary transport or both. For example, a tree length harvesting system is one in which the tree is felled and is partially processed by removing the top and limbs before primary transport. After the stem is transported to the landing the stem is generally processed further, into log lengths, before secondary transport.

Felling or cutting is normally the first step in timber harvesting and greatly influences the productivity of other harvesting functions that follow. Felling operations should be undertaken in ways that benefit later operations, in particular primary

J. Wang (✉) • D. Hartley • W. Liu
Division of Forestry and Natural Resources, West Virginia University,
322 Percival Hall, Morgantown, WV 26506, USA
e-mail: Jxwang@wvu.edu

M. Jacobson and D. Ciolkosz (eds.), *Wood-Based Energy in the Northern Forests*,
DOI 10.1007/978-1-4614-9478-2_5,

Table 5.1 Capability of example extraction methods

Method and example equipment	Power (hp)	Payload (m^3)	Skidding distance (m)	Productivity (m^3/h)
Animal (draft horse)	–	1.1	152	3.6–7.2
Crawler (JD 450)	80	1.4–2.8	152	10.9–21.7
Farm tractor (log hog)	40	1.1	152	7.2–10.9
Rubber-tired skidder (JD 640)	130	3.6	457	18.1–43.4
Track-laying skidder (FMC 210)	200	5.4	305	18.1–43.4
Forwarder (Gafner 5510)	100	10.9	610	18.1–43.4

Note: Brands listed here do not imply endorsement and are merely an example. Source: Greene and Reisinger (1999)

transport. Felling can be extremely hazardous to both the feller and others working on the ground of a harvesting operation. Increased mechanization of felling operations has been able to reduce the risk but not completely eliminate it.

After felling the severed stems must be moved to a central location to improve the efficiency of the following processes. The process of moving cut wood from the stump area to a "landing" or a staging ground where wood is collected prior to secondary transport (trucking to the mill) is called "primary transport." The vast majority of wood that is extracted throughout the eastern United States, and for that matter the world, uses ground-based methods for primary transport. In most cases, ground-based harvesting is the most economical system but is sensitive to topographical factors that can affect the safety, productivity, and cost of the operation. Conditions such as steep or rugged ground, wetlands, or environmentally sensitive areas are problems for ground-based systems and generally require the use of an alternative extraction method.

Ground-based systems can range from the inexpensive and primitive to the capital intensive and technologically advanced (Wang and Greene 1999). The six basic types of ground-based systems are listed with their production capabilities (Table 5.1). Skidding is the process whereby logs are dragged on the ground from the stump to the landing. While animals are still used in some applications, since the mid-twentieth century the majority of skidding has been completed with the use of machinery. Machine development has led to many different configurations for both tracked and rubber-tired carriers.

The major difference between forwarders and skidders is that forwarders carry their payload completely off the ground, instead of dragging it along the ground. Since the entire weight of the load is carried by the machine, forwarders are generally limited to transporting log length material or pulpwood or woody biomass. Also, carrying the entire load on the machine causes forwarders to exert high amounts of pressure on the ground; for this reason forwarders are generally operated on soils that have high soil strength or when using designated skid trails (Greene and Reisinger 1999).

Once the stems have been moved to the central location, they must be processed to improve the efficiency with which they can be handled and transported by truck and/or rail. Processing includes delimbing and/or bucking or chipping or grinding. In addition to improving the transportation and handling characteristics, the objectives of processing are to (1) produce the wood or the biomass products with specifications desired by the marketplace with acceptable quality and (2) minimize fiber/mass or value loss, all at an acceptable per-unit cost. Product types depend on species being processed and the end product specification. For example, the most common products delivered to markets in the Northeast are logs. Basically, three types of forest products are produced in the region: tree-length trunks, specified-length logs, and chips.

Upon completion of processing, the material must be loaded for transport to a mill. Loading is the last operation that completes the series of tree processing activities before transportation to the mills. It involves picking up forest products (logs, slash bundles, chips) into a log transportation vehicle (log truck) or a chip van. Loading of log is usually done at the landing and seldom done at the stump area or the reload point. It is a simple operation which can be handled with a host of equipment from handheld tools to hydraulic controlled machines.

Today truck delivery of roundwood or biomass is by far the most flexible transportation method and the most widely used. Trucks vary in load carrying capacity. Two generally used trucks for log transportation are straight trucks and tractor-trailers. In addition, chip vans are used to hold chips produced in the woods or at a mill facility. The efficiency of loading and trucking could be affected by factors such as size and type of products, hauling distance, payload size, and turnaround time (Wang 2007).

Harvesting systems can be classified by any number of criteria. Typically, they are classified by their level of mechanization, the equipment used, or products produced. A harvesting system that consists of chainsaw felling and cable skidder for primary transport could be classified as any of the following: partially mechanized, ground-based, motor-manual, or skidder. Additionally, depending on the products being produced the same system could be classified as a short wood, log length, tree length, whole tree, or chips. Classification of harvesting systems is generally used for comparison and to differentiate between two systems, i.e., partially mechanized versus fully mechanized, ground based versus cable or aerial, and log length versus tree length. Throughout the remainder of this chapter the basic differentiation between systems will be made based on the product being harvested, i.e., biomass residues or short-rotation woody crops (SRWC).

5.2 Harvest System Analysis

5.2.1 Systems for Biomass Residues

Woody biomass can be removed from the forest using different methods. Transporting loose residue, chipping at the landing, and bundling are all viable options (Spinelli et al. 2007). For example, slash bundling or baling systems can

Fig. 5.1 Wood energy bundler (photo courtesy John Deere Corp.)

Table 5.2 Sample productivity and cost of selected residue harvesting systems

System	System configuration: Machine	No. of machines	Productivity (green tons/SMH): Machine	Productivity (green tons/SMH): System	Cost ($/SMH): Machine	Cost ($/SMH): System	Unit cost ($/green ton)
Cable skidder-loose residue	Cable skidder	2	6.42	12.06	52.87	105.74	8.77
	Loader	1	12.06	12.06	55.87	55.87	4.63
Cable skidder-chip	Cable skidder	3	6.42	19.25	52.87	158.61	8.24
	Chipper	1	28.72	19.25	139.21	139.21	7.23
Grapple skidder-loose residue	Grapple skidder	1	10.56	10.56	66.77	66.77	6.33
	Loader	1	12.06	10.56	55.87	55.87	5.29
Grapple skidder-chip	Grapple skidder	2	10.56	21.11	66.77	133.54	6.33
	Chipper	1	28.72	21.11	139.21	139.21	6.59
Slash bundler	Slash bundler	1	9.13	9.13	130.33	130.33	14.28
	Forwarder	1	17.41	9.13	94.34	94.34	10.34

Source: Wu et al. (2011)

promote full-tree utilization, reduce hazardous fuels, reduce logging site preparation costs, eliminate in-woods chipping, and provide long-term storage over chips (Fig. 5.1). The collection of woody biomass can be either accomplished after harvesting activities or incorporated into the harvest of roundwood or other products; as a result, the overwhelming majority of residue harvesting systems are based on conventional harvesting systems with the potential addition of chipping or grinding equipment. Woody biomass, in the form of branches, tops, etc., can be extracted to landings by a skidder or forwarder (Wang et al. 2004a, b, 2005; Li et al. 2006; Grushecky et al. 2007), and the loose residue can then be delivered to consumers via truck or truck-and-trailer units (Table 5.2). This system is constrained by the difficulty of fully utilizing vehicle payload and is preferable only over a short hauling distance (Spinelli et al. 2007). The process of chipping or grinding woody material can improve bulk density, homogeneity, and handling characteristics of the raw material from the forest (Johansson et al. 2006). In-woods chipping as part of conventional harvesting or thinning has typically been the most cost-effective system to recover forest residue for biomass (Hartsough et al. 1997; Jeuck and Duncan 2009), which relies on the smooth functioning of all steps in the woody biomass supply chain. Downtime is costly when a delay occurs at the chipper, on the truck, or at the utilization plant (Jackson et al. 2010). Bundling is a possible alternative to improve forest fuel logistics (Spinelli et al. 2007). Biomass bundlers collect, compress, and bind forest residue into cylindrical bundles (Fig. 5.1). Processing speeds for bundlers are typically in the range of 10–30 bundles of biomass per hour. A standard bundle is approximately 3.5 m in length, about 0.8 m diameter, and weighs 450 kg. With adequate compression, bundles can allow full trucking payloads and make handling such as loading and processing more efficient (Rummer et al. 2004).

5.2.2 Systems for Short-Rotation Woody Crops (SRWC)

Throughout the United States, the harvest of SRWC has historically been done using conventional harvesting systems, similar to the systems that have been discussed in the previous section. In order to be able to produce at levels that would produce feedstocks at sufficiently low price to be beneficial for energy production, these systems are almost entirely highly mechanized. While these types of equipment are able to have very high production in certain conditions, they have not been designed to work optimally in short-rotation environments and thus are probably not the ideal choice of equipment into the future.

The need of new equipment and harvesting systems has been realized by both the academic community as well as equipment manufacturers. As a result, several paths have been explored to develop harvesting systems capable of handling SRWC efficiently.

The two main paths that have shown the most promise can be classified as cut only and cut and chip or cut and bale (Table 5.3). Cut-only machines only fell the feedstock with densification and/or comminution taking place in later phases, while the cut and chip-type systems combine the phases of felling and comminution into a single phase. Through early trials, indications are that the cut and chip harvesters are the most promising types of harvesting systems for SRWC, because the material that is produced has desirable handling characteristics (i.e., chips vs. full trees) and downtime is reduced as a result of less machine interaction than in a cut-only system.

The trend has been to build the harvesting platforms on historically agricultural machine platforms rather than design-purpose-built machines. This makes sense since the small diameters of SRWC, which is utilized for energy, are similar in characteristics to traditional agricultural crops such as corn or sugarcane (Hartsough et al. 1996). One example of such a machine is the Case New Holland Coppice Header that is being developed in conjunction with the State University of New

Table 5.3 Systems for harvesting short-rotation woody crops

Manufacturer	Machine	Type	Drive	Max. material size (cm)	Source
Anderson	WB-55	Cut and bale	PTO	15	1
FECON	Bio-Harvester	Cut and chip	PTO	15–20	2
GyroTrac	BBS 1000	Cut and bale	Autonomous	No limit	3
New Holland	130FB	Cut and chip	Autonomous	15	4

1—http://www.grpanderson.com/en/biomass/biobaler-wb55
2—http://www.fecon.com/images/pdf/literature/products/9/Bio-Harvester-6.pdf
3—http://www.gyrotrac.net/2011/06/bio-energy-baling-system/
4—http://www.thecleanenergyleader.com/en/biomass_en/FarmingTechnology/Forageharvesters.html

Fig. 5.2 SRWC coppice header (source: Abrahamson et al. 2010)

York College of Environmental Science and Forestry (Fig. 5.2). It is one of the first dedicated harvesters for SRWC. That harvester can handle materials of up to 15 cm with a production rate of 1–2 ha per hour (Abrahamson et al. 2010).

5.3 Determining Harvest Productivity and Cost

5.3.1 Measuring Harvest Productivity

Harvesting machine productivity can be defined as a measure of log or chip volume/weight produced from a given forest site over a given time. Productivity measured in volume or weight per unit time usually increases with increases in material piece size. Time and motion study is one of the most popular ways used for examining harvesting machines' production rates. Throughout the years, time and motion studies have been conducted in different ways (Wang et al. 2003). Early in the twentieth century, stopwatches and paper were used to measure and record times (Howard 1989). This method usually required two people working together. One operated the stopwatch and recorded the times, and the other took measurements about the site and volume of timber or biomass being removed from the forest. This method is probably the most common method used in logging productions.

However, such traditional time studies can be very tedious, expensive, and error prone (Olsen and Kellogg 1983) and are being replaced by methods that utilize modern technologies such as video, handheld computers, and Global Positioning Systems (GPS) (Howard and Gasson 1991). The use of video to aid in the collection of time study data has the benefit that it captures exactly what is happening while the time measurement is recorded. Information about the site or the volume of timber or biomass is still needed, so extra time and resources are still required to collect these data.

Additionally, with the proliferation of handheld computer technology, new techniques have been introduced into computer-based time studies of forest operations or woody biomass harvesting. Handheld time study systems have the advantages of being able to modify the time design "on the fly" and the economy of keystrokes (Howard and Gasson 1991). Designing, editing, and preparation of the time study are traditionally done on a base computer, and the edited data are saved in a format that can be easily transferred between the base and handheld unit. Another benefit of these systems is that they use the built-in clock of the handheld computer to keep track of the elemental time. This allows the observer to use the program to collect times while concurrently entering the site and volume or mass weight information.

The most recent advancements in the collection of time study data are the use of GPS technology to automate the process. An automated time study of felling and skidding was developed by McDonald and others (McDonald 1999; McDonald and Rummer 2000). A GPS was used for tracking machine movements and switches for monitoring machine functions. The system was successful in providing gross time study data but less so in providing detailed elemental times. Their results also indicated that the sequencing of tree cuts with felling cycles was subject to errors. Other time study systems have been developed as well that tend to incorporate various data collection and processing approaches.

5.3.2 *Harvest System Cost Analysis*

Harvesting systems and methods used for harvesting activities greatly affect overall harvesting cost, productivity (volume produced per hour during a harvesting operation), overall profitability of harvesting operations, returns to a landowner, and ecological impact (Han et al. 2004). Today, there exists a spectrum of harvesting equipment ranging from a simple chainsaw to complex purpose-designed equipment that can fell, delimb, buck, chip, and haul logs or chips. To select specific equipment from this spectrum and use profitably, it is imperative that harvest operators know how to estimate costs and productivity of the equipment in use (Miyata 1980; Greene and Lanford 2001). A variety of factors contribute to biomass harvest planning difficulties, including harvesting equipment allocation, terrain conditions, tract size, inflation, and labor costs (Miyata and Steinlulb 1981). Each of these factors can be modeled and improved using harvesting system cost analysis tools.

During the last 50 years, the two most popular logging cost analysis methods that have been widely applied in the field include the machine rate and cash flow approaches. Matthews (1942) detailed cost factors in the logging industry and presented a method for unit cost and machine rate calculation, which was considered as one of the earliest attempts at logging cost analysis. Based on Matthews' method, Miyata (1980) developed a procedure to estimate logging equipment costs. This method, now referred to as the "machine rate" method, is included in many current computer-based logging business analysis programs. The machine rate method combines fixed cost of machine ownership and variable cost of operation to calculate a machine's lifetime average hourly cost (Brinker et al. 2002). While the machine rate method is the most commonly used method, for cost estimation, because of its relative ease of calculation; application requires caution on the part of the manager. Caution is required because the result is the average cost of ownership and operation over the life of the machine and may not accurately reflect the costs that are experienced at a specific point in time. Actual ownership costs will be greater than the machine rate early in the machine's life and will change as the machine ages (Brinker et al 2002). Machine rate is also affected by other basic assumptions such as total hours of machine life, change in time value of money (inflation), interest rate, insurance, and taxes (Tufts and Mills 1982; Brinker et al. 2002). Using the machine rate, the major components of machine cost are the following:

1. Ownership (fixed) cost is the total cost accumulated on a machine over the economic life of the machine. Economic life of a machine is the period (years) for which a machine can work profitably. The components of ownership cost are depreciation and interest rate, insurance, and tax.
2. Operating cost (variable cost) includes cost of fuel, lubrication, maintenance and repair (M&R), tires, and labor. This cost varies with output level. Fuel and lube cost is related to machine horse power and transmission. Lube cost includes the cost of engine oil, hydraulic oil, and lubricants. It is usually determined as percentage of fuel costs.
3. Labor cost is the hourly wages paid to worker for the entire daily shift to produce timber in the forest. Part of this component may include contribution to social security, unemployment insurances, and workers' compensations. In machine rate calculations, labor cost is calculated by multiplying wage rate by a fraction for fringe benefits. Fringe benefit percentage can be obtained from payroll records or rough estimates.

The machine rate only addresses the direct costs associated with harvesting such as the costs of felling, skidding, chipping, loading, and hauling through the ownership and operations of the machine. While these costs are important and usually command most of a contractor's attention, accounting for 65–85 % of the total cost of a harvesting operation, it is important not to lose sight of the remaining 15–35 % of the total cost (Greene and Reisinger 1999). These remaining cost factors associated with harvesting are often labeled as "indirect" costs, and lack of attention to these cost factors can also result in an operational loss. Examples of indirect cost are as follows: moving expenses from one tract to another, permits, road and landing construction, and overhead expenses.

The machine rate method presented in the previous section is appropriate when the desire is to ascertain the cost of operating a single machine without concern for the interactions that occur within a harvesting system. In a harvest system, there may be multiple machines performing a single function. For example, two skidders pull behind one feller-buncher or a single biomass harvester serves two chip vans. Each function has inherent productivity under given conditions. This is to say, a machine taken by itself will have no restraints from its sister functions. It will produce biomass at some rate plus or minus its inherent variability.

When the machines are added into a harvesting system it is often found that the harvesting machines do not always produce at the same rate. Some are more productive than others, and some have higher utilization level than others. In this case the productivity of the system is then limited to the production of the least productive piece of equipment or function. The least productive function in a harvest system is called the limiting function. To mesh functions together effectively to form an efficient harvest system, each function must produce biomass at about the same rate. The process of making each function in a system produce equally is called system balancing. There are four ways to balance an operation: (1) add or subtract machines, (2) work some functions in more or less scheduled hours, (3) change work conditions, or (4) change machines (Wang et al. 1998; Greene and Reisinger 1999).

5.4 Economic Modeling of Biomass Harvest

5.4.1 Modeling Process

Most recently the analysis of harvesting operations has moved away from the examination of single functions or systems to a more general modeling of the entire supply chain to form a more complete picture of the operations and the importance that the interactions between functions have in the determination of the final delivered cost of a feedstock. A supply chain for woody biomass consists of these main components: feedstock establishment, cutting, extraction, processing, bundling (if any), storage, loading and hauling, and chipping/grinding under different feedstock and operational conditions (Fig. 5.3) (Wu et al. 2011). Since the development of a bioenergy industry will depend on the cost-effectiveness of how biomass can be harvested, transported, stored, and processed, economic modeling of biomass harvest and logistics may offer the best approach to optimize economic scenarios of biomass utilization for bioenergy.

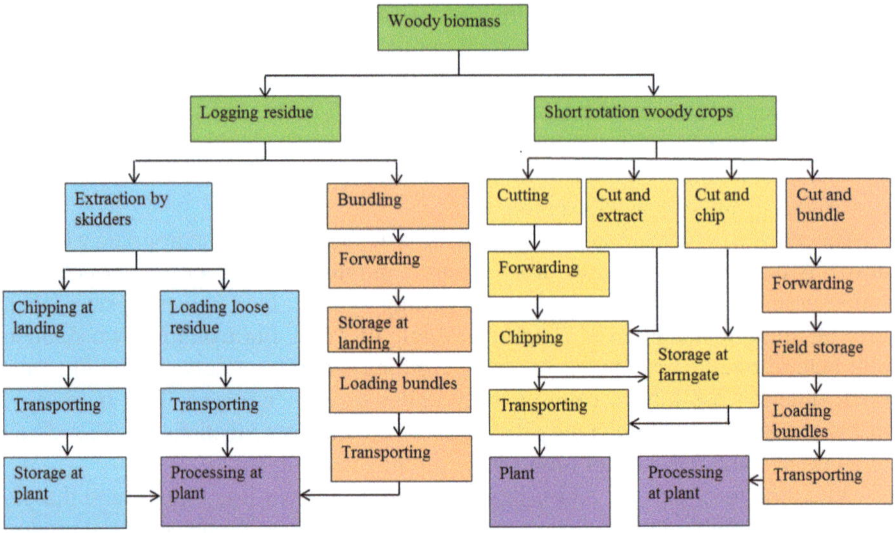

Fig. 5.3 A modeling framework of biomass harvest

The following section presents an example of a situation in which economic modeling was utilized to analyze the harvesting and extraction of forest biomass on a large scale to optimize a woody biomass-based biofuel facility's location with the objective of minimizing the total annual delivered cost of woody biomass under resource and operational constraints (Wu et al. 2011).

5.4.2 *Modeling Case Analysis*

A mixed integer programming (MIP) model was developed and solved using General Algebraic Modeling System (GAMS)/CPLEX, and the model was applied in the central Appalachian region within the state of West Virginia. Five woody biomass handling systems that perform bundling (if any), extraction, storage, hauling, and/or chipping/grinding activities were included in the model. These systems are cable skidder-loose residue (S1), cable skidder-chips (S2), grapple skidder-loose residue (S3), grapple skidder-chips (S4), and forwarder-bundle (S5). Systems S1, S3, and S5 used cable skidders, grapple skidders, or forwarders to collect and transport scattered logging residues from stump to landings, respectively, and a wheeled log loader was used to load residues onto a tractor-trailer for hauling; then the grinding was processed at the utilization plant. Similarly, systems S2 and S4 used the above extraction machines for residue collection. Then logging residue was chipped using self-loading chippers at landings, and chip vans were used to deliver the chips to the utilization plant. System S5 used a slash bundler to bundle logging residues at the stump, and a forwarder was used to forward the bundles to landings and pile up the

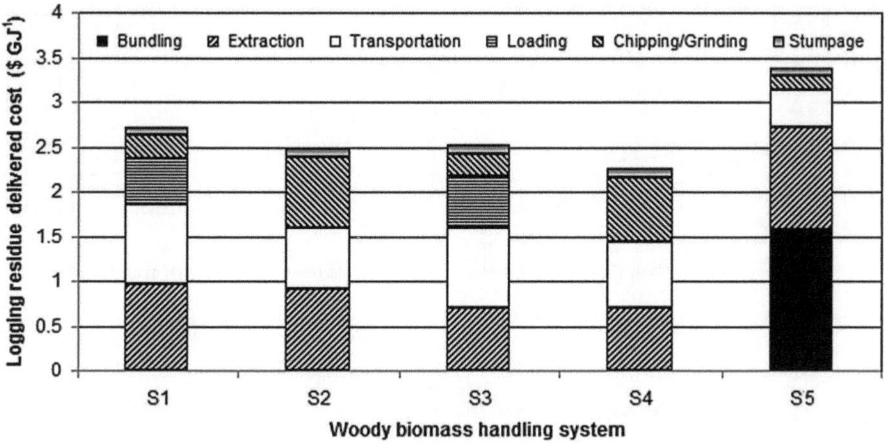

Fig. 5.4 Delivered cost composition of woody biomass by handling system for the base case. S1—cable skidder-loose residue, S2—cable skidder-chips, S3—grapple skidder-loose residue, S4—grapple skidder-chips, and S5—forwarder-bundle (source: Wu et al. 2011)

bundles. After storing for a period of time for drying, the bundles were transported to the utilization plants where the grinding process was accomplished.

5.4.2.1 Delivered Cost of Woody Biomass

The delivered cost of logging residue included the costs associated with stumpage, bundling (if any), extraction, storage (if any), loading and hauling, and chipping/grinding. For a base-case scenario of daily demand of 900 metric tons of dry woody biomass, the average delivered cost ranged from \$2.30 GJ^{-1} to \$3.02 GJ^{-1} (\$36.54 to \$53.75/dry ton). Extraction and transportation were the major cost components for systems S1–S4, accounting for an average of 33 and 32 % of the total delivered cost, respectively, which were followed by chipping/grinding and loading (Fig. 5.4). For the forwarder-bundle system (S5), bundling was the major cost component and accounted for 46 % of the total delivered cost. It was followed by extraction (34 %), transportation (12 %), grinding (5 %), and stumpage (3 %). The comparisons among the handling systems for the base case indicated that delivering chips was slightly cheaper than shipping loose residues to the optimal plant because the cost savings associated with trucking using a chipping system were more than the offset of centralized chipping at the utilization plant using a loose residue-based system. No direct storage cost was incurred in any of the systems. Since timber harvesting takes place year round in West Virginia, logging residue is accordingly available and ready for collection for the whole year. Typically, there is no need to store logging residue at landings or roadside. Onsite storage incurs in system S5 because slash bundles extracted were assumed to be entered into storage for drying. However,

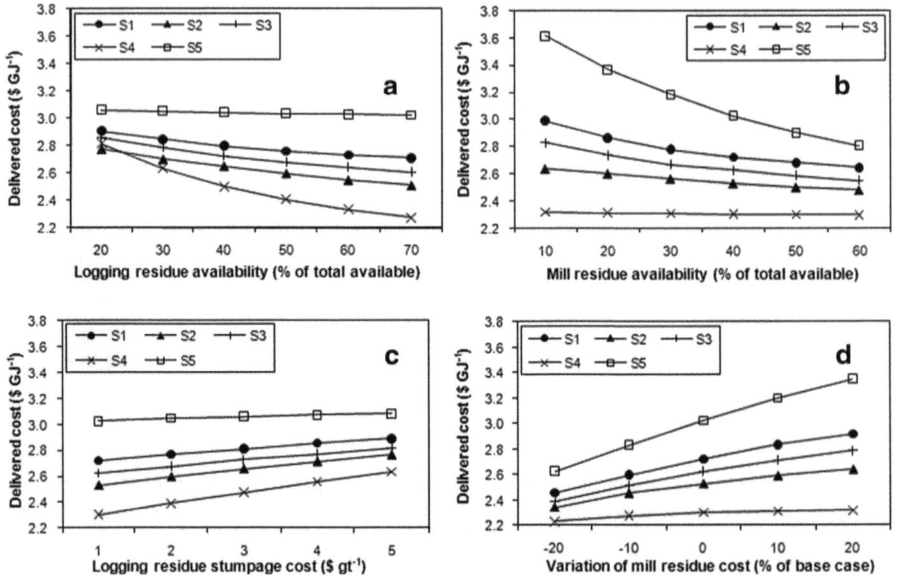

Fig. 5.5 Sensitivity analyses of the woody biomass delivered cost by residue availability and purchase cost (1 gt = 1 green metric ton) (source: Wu et al. 2011)

storage cost due to the time-value-of-money concept related to skidding/collection costs incurred was not considered here.

5.4.2.2 Sensitivity Analyses

Extensive sensitivity analysis was performed under different scenarios, including biomass availability and purchase/stumpage price, demand, extraction distance, and fuel pricing. The delivered cost was mostly affected by woody biomass demand. Extraction distance had the least impact on the delivered cost.

Markets for biomass and willingness of forest landowners to sell can determine whether or not biomass should be removed from the forest at all (Bowe and Bumgardner 2006; Becker et al. 2009). A decrease in logging residue availability would increase the delivered cost of woody biomass due primarily to longer hauling distances (Fig. 5.5a). Compared to the initial case where 65 % of logging residue is available, if the availability of logging residue was down to 30 %, the delivered cost would increase 4 % ($0.13 GJ^{-1}) for the cable skidder-loose residue system, 6 % ($0.17 GJ^{-1}) for the cable skidder-chips system, 6 % ($0.16 GJ^{-1}) for the grapple skidder-loose residue system, 13 % ($0.33 GJ^{-1}) for the grapple skidder-chips system, and 1 % ($0.03 GJ^{-1}) for the forwarder-bundle system. Efficient harvesting such as collecting logging residue during timber harvesting is critical to the future economic viability of logging residue utilization (Grushecky et al. 2007).

However, the implications of increased harvest efficiency must be weighed against potential impacts on site productivity, especially for low-productivity sites (Scott and Dean 2006).

As a by-product, mill residue is limited by production constraints in sawmills. The utilization of mill residue for biofuels will compete with other established uses such as pellet fuel, boiler fuel, and others in the central Appalachian region. A shortage of mill residue from sawmills that have closed because of poor lumber markets may result in the escalation of the prices of forest-based biomass (Becker et al. 2009). If the mill residue availability changed from 40 % (base case) to 20 % due to other needs of mill residues, the delivered cost would increase 5 % for the cable skidder-loose residue system (S1), 3 % for the cable skidder-chips system (S2), 4 % for the grapple skidder-loose residue system (S3), 0.4 % for the grapple skidder-chips (S4), and 11 % for the forwarder-bundle system (S5) (Fig. 5.5b). This increase in delivered cost would be due to an increase in the portion of the raw material requirements that would need to be made of forest residues, which would in turn expand the collection areas, which would increase the transportation costs.

In the current market, the value of uncollected logging residue is very low. Therefore, a stumpage price of \$1 metric ton^{-1} was assumed for logging residue in the base case. As the market for biomass develops, more biomass will be needed, and forest landowners may expect a higher return for these raw materials. The delivered cost would increase \$0.01–\$0.09 GJ^{-1} as a result of every 1 dollar per green ton increase in the stumpage cost of logging residue (Fig. 5.5c). The assumed purchase price of mill residue was based on current market pricing in the region. Mill residue has been traditionally used as pellet and boiler fuel in the region. The potential competition between current uses and prospective utilization for biofuels could raise the price of mill residue, especially when the demand is high. Sensitivity analysis indicated how the variation of mill residue cost affected the delivered cost of mill residue (Fig. 5.5d). Increasing the purchase price of mill residue by 10 % would cause an increase of delivered cost by 4 % for the loose residue-based systems (S1 and S3), 1 % for the chips-based systems (S2 and S4), and 6 % for the forwarder-bundle system (S5). This again is due to the increase in demand for mill residue and the need to travel further to acquire sufficient amounts. Therefore, the delivered cost was much more sensitive to the purchase price and availability of mill residue than logging residue when using systems S1, S3, and S5.

The average delivered cost of woody biomass for the base case ranged from \$2.30 GJ^{-1} to \$3.02 GJ^{-1} using the different woody biomass handling systems. This estimation was within the range of \$1.55–\$5.09 GJ^{-1} (\$27–\$89 dry metric ton^{-1}) estimated by Grushecky et al. (2007) for the same region. However, it was higher than the US Department of Energy (USDOE) target cost of \$2.21 GJ^{-1} (\$38.58 dry metric ton^{-1}) for 2012 at which level the production cost of biofuels (ethanol) from woody biomass could be competitive with gasoline (Perlack et al. 2005). Several factors contribute to the higher delivered cost of woody biomass in the central Appalachian region, including the mountainous terrain, hardwood species, and fuel economy. Cable and grapple skidders are the most commonly used extraction machines partially due to the difficult terrain in the region. The fuel economy for hauling woody biomass in the region was estimated at 2.12 km L^{-1} (5.0 miles $gallon^{-1}$) after consulting with local loggers; however, it could be up to 3.83 km L^{-1} (9.0 miles $gallon^{-1}$) in other regions with gentle terrain (Jensen et al. 2002). The delivered cost could be reduced if extracting logging residue during harvest (Grushecky et al. 2007). The main reason for this is that the machinery cost will be distributed over multiple products in this type of system. The cost of extraction for the residues will then be effectively lowered as a result. In addition, some of the costs will be avoided as the machinery will not need to be moved in and out an additional time.

More importantly, some of the cost estimate parameters were based on reasonable assumptions from literature and are subject to change. For instance, a large-scale woody biomass-based biofuel facility could increase demand competition for woody biomass and raise the price of raw materials. Meanwhile, as demand for woody biomass increases, more loggers may engage in the biomass harvesting and processing business or expand their operations. Currently, fewer logging companies are involved in woody biomass harvesting and processing. The infrastructure needed to provide woody biomass to plants will have to be developed as markets come online (Becker et al. 2009; Missouri Forest Products Association 2010). Fuel price can fluctuate with worldwide and regional supply and demand, seasonal demand, and other factors. It is also worthy of note that the delivered costs in this study primarily focus on the direct costs associated with biomass handling activities. These costs are important and usually command most of a contractor's attention. There may be other incurred costs such as moving expenses if biomass collection is accomplished in a separate activity. These indirect costs could be considered in future modeling efforts.

5.5 Tools for Harvest System Analysis

Traditionally, harvesting analysis systems were programmed with either a simple spreadsheet without detailed business functions or with a programming language that is too complicated for field operators to understand and apply into their operations. Several programs have been developed for improving harvesting production/cost analyses including the Auburn Harvesting Analyzer, Harvesting System Analyzer, and Harvesting System Simulator (Wang and LeDoux 2003; Hendricks 1986; Stuart 1981; Tufts et al. 1985). Reisinger and Gallagher (2001) compared these three programs in terms of input and output performance. The Auburn Harvesting Analyzer is a spreadsheet-based program and relatively easy to use. However, it was developed for use with southern logging and stand conditions that can be very different than those in mixed hardwood forests. The Harvesting System Analyzer and Harvesting System Simulator both work well for research purposes, but they can be too complicated for many loggers to use efficiently.

Further refinement of these methods was implemented by Greene and Lanford (1996) who described a step-by-step logging cost analysis method that includes machine time and productivity, regression models for estimating production and model selection, estimating costs of equipment, and analysis of an entire logging system. Visser et al. (2004) developed a program which is capable of predicting production and cost of the harvesting system under certain stand and terrain conditions. However, the user must input the production potential as well as fixed and variable cost estimates for each machine involved. A software package titled RILSIM, designed for financial analysis of low-impact and traditional logging systems, was released by Dykstra (2004). The application of RILSIM has been somewhat limited as detailed wage, equipment, and activity data required for analysis is often not available. Further attempts have been made based on a harvesting simulator developed by Wang and Greene (1999), in which a numerical forest harvesting simulator was developed for users to perform harvesting production/cost analysis in central Appalachia (Li 2005). Again, this numerical simulator's limitation is that it is still too complicated for many loggers to efficiently use in their businesses.

Models and tools have been developed for biomass harvest analysis. A list of models that can be useful in the analysis of biomass harvesting systems is presented in Table 5.4, including the names of the models with a short description as well as needed inputs and outputs.

The crop budget model EcoWillow was developed by State University of New York College of Environmental Science and Forestry for Short Rotation Woody Crops (Buchholz and Volk 2011). It is built with Microsoft Excel and contains four input/output sheets and four output graph sheets. This model contains three sub-models: planting, harvest, and transportation costs. It allows the user to input site-specified variables and to calculate cash flow and profits for the wood chips in the entire logistic chain.

The Policy Analysis System (POLYSYS) is developed by the University of Tennessee's Agricultural Policy Analysis Center and USDA's Economic Research Service and the Oklahoma State University's Great Plains Agricultural Policy Center (De La Torre Ugarte and Ray 2000). It is a modular economic simulation modeling system of the US agriculture sector where crop demand and price are estimated and policy decisions are made based on the simulated results. POLYSYS works as an umbrella framework. It contains four modules: regional crop supply module, national livestock module, national crop demand module, and national income module. The four modules also interact with each other. This model is useful for biomass- and bioenergy-related applications. It has been applied in many places, but there are some additions needed if non-typical crop and livestock commodities are considered.

The Integrated Biomass Supply Analysis and Logistics (IBSAL) model, which is developed by Oak Ridge National Laboratory, is designed to simulate the collection, storage, and transport operations for supplying agricultural biomass to a biorefinery and to predict the delivered costs (Sokhansanj et al. 2006). This model contains a network of independent modules and connectors which are included in the supply chain. The model considers biomass availability, moisture content of biomass, weather factor effects, equipment performance, transport equipment performance, and dry matter loss. This model requires some input data such as number of farms involved, average yield, harvest schedule, and moisture content of the biomass. Daily weather data are also needed. Equipment information is necessary to be

Table 5.4 Available models for biomass harvest analysis

Model	Inputs	Outputs
EcoWillow Buchholz and Volk (2011) Excel-based model to calculate cash flow for the establishment and harvest of willow for bioenergy	Desired rotation length Expected crop yield Cost of land rent Insurance Crop establishment costs Subsidies (optional)	Cash flow diagrams Cost distributions Yearly cash flows Accumulated cash flow
Polysys De La Torre Ugarte and Ray (2000) A modular economic simulation model of the US agriculture sector, to solve for demand and price of agronomic crops at the national level	US Agriculture baselines Crop rotations Regional soil info Land use information Cost and return data for crops and livestock	Harvestable acres Export and export value Average crop prices Net return
IBSAL Sokhansanj et al. (2006) A dynamic biomass supply and logistics model that simulates the collection, storage, and transport of biofeedstocks to a biorefinery	Spatial information Schedule Daily yield Machine production and cost Weather info	Quantity of biomass delivered Operation length Delivered biomass properties Collection and transportation costs Energy inputs Carbon emissions
WTRANS Jensen et al. (2002) A spreadsheet based model that estimates the transportation costs and cost difference between wood residue types	Truck weight laws Truck volume limits Speed limits Operation costs Biomass density Biomass price	Delivered price per ton Fuel use comparison of transportation options
Woody biomass harvest and logistics analysis system Wu et al. (2011, 2012) A mixed integer linear programming model to estimate the delivered price of woody biomass to a conversion facility	Woody biomass availability Machine operation constraints Machine productivity Machine costs	Quantity of biomass delivered per year and price Cost components of harvest and delivery Ideal plant locations

provided to calculate service time and cost. While both POLYSYS and IBSAL have been developed as agricultural models, the framework that they provide could be easily adapted and modified for use in the areas of woody biomass.

The Wood Resource Transportation Analysis System (WTRANS) is a spreadsheet-based program, developed by the University of Tennessee (Jensen et al. 2002). It can be used to examine where hardwood and softwood green mill residues may be located on a county-by-county basis in Tennessee and estimate the transportation and delivered product costs, potential residue needs for energy use, and fuel cost comparisons using residues for greenhouse heating. Additionally, the INRS Diesel and Trucking Calculator provides another method to examine the impact that constantly changing diesel prices and truck configurations have on the price of biomass feedstocks (Kingsley 2011).

A Woody Biomass Harvest and Logistics Analysis System has been developed by West Virginia University to analyze harvest using traditional forest harvesting machinery (Wu et al. 2011, 2012). It was developed for forest engineers, foresters, harvest contractors, and biomass managers to assist the analysis of forest/biomass harvest and logistics. This system uses a number of associated models from previous studies that provide a solid contribution to harvest cost estimation and decision support. Compared with the previous spreadsheets or computer programs, this system offers a user-friendly and interactive interface and provides additional capabilities such as machine rate information, system balancing, and a summary presentation. In addition, multiple simulation results based on different scenarios (e.g., extraction distance and turn payload) can be conducted and saved to allow users to select a harvesting system that best achieves desired results.

The system could be very useful for harvesting management decision-making, including harvesting system configuration, equipment productivity and cost estimation for each harvesting function, determining the optimum mix of equipment for a balanced system, system productivity and cost estimation, and system performance and cost efficiency evaluation under varied stand, site, and machine conditions. It could also be helpful in preparing bids for timber or biomass sales by estimating the costs for different harvesting systems, harvest methods, and road scenarios. To apply the system in other regions, adjustments may be needed to account for differences in site, stand, weather, or other operational conditions. An optimization component will be integrated into the system to estimate the delivered cost of woody biomass in a supply chain and optimize the supply chain scenarios.

5.6 Summary

Harvest is one of the most costly components of the delivered price of wood feedstock to a conversion facility. Being able to determine which harvest systems will be the most financially and environmentally effective means of extracting the feedstock is paramount to the ultimate success of a bioenergy/biofuel industry. This means that the individuals or the groups that are tasked with supplying the feedstock must be able to accurately determine the appropriate harvest methods and understand the trade-offs between methods.

The analysis of harvesting operations is typically based on the analysis of harvesting systems, as it is often more important to understand the interactions of each piece of equipment in the system with their neighboring functions than to understand the capability of each piece of equipment on its own. Keeping this in mind, the methods that are used to analyze the productivity and cost of harvesting operations focus on key operations that are performed, irrespective of the equipment mix or the feedstock that is being harvested. In all cases the harvest operation will have the key components of felling, primary transportation, processing, loading, and secondary transportation. While it is true that in some harvesting systems it is possible that some functions will be combined with other functions, they will nonetheless be present.

This is not to say that the feedstock that is being harvested or the conditions in which the systems are operating are not important, quite the contrary. Understanding the design and capabilities of each machine in the system that you are proposing can eliminate the selection of components that will result in systems that are not cost effective or simply unable to effectively work in the selected environment. For example, if forest residues are the selected feedstock, it is most likely that the correct equipment mix will contain primarily conventional timber harvesting equipment with the addition of some specialty equipment to handle and process the biomass for transport. This equipment has been shown to be effective for this type of harvest and can attain production levels required to produce feedstock at an acceptable price point, with only minor changes to traditional work flows. Conversely, the harvest of SRWC will require the use of highly mechanized and often purpose-built machinery to accomplish the task economically.

Designing the optimal harvesting system for a given stand or planting is not a trivial task, and while it is easy to discount some options based on terrain, material size, slope, etc., there will oftentimes be multiple choices that are seemingly appropriate. In these cases, it is necessary to perform some analysis beforehand to determine the system that should be employed. Fortunately, there are several methods and tools that are available to perform such tasks. The method of analysis and tools that should be chosen will be defined by the scope and level of detail that is needed for the analysis, i.e., analyzing a system for a single harvest site to predetermine the cost or the evaluation of a harvest system or a group of harvest systems for a regional supply scenario. In the former, the use of a simple spreadsheet or even paper calculations may be appropriate while the latter will require the use of not only mathematical models but also potentially geographic information systems as well as other tools.

References

Abrahamson L, Volk T, Castellano P, Foster C, Posselius J (2010) Development of a harvesting system for short rotation willow and hybrid poplar biomass crops. SRWCOWG, Syracuse, NY, October 18, 2010

Becker DR, Abbas D, Halvorsen KE, Jakes PJ, McCaffrey SM, Moseley C (2009) Conventional wisdoms of woody biomass utilization. http://www.forestguild.org/biomass/resources/ISE_Biomass.pdf. Accessed 18 May 2010

Bowe SA, Bumgardner MS (2006) Small-diameter timber utilization in Wisconsin: a case study of four counties. N J Appl Forest 23(4):250–256

Brinker RW, Kinard J, Rummer B, Lanford B (2002) Machine rate for selected forest harvesting machines. Alabama Agricultural Experimental Station, circular 296, p 12

Buchholz T, Volk T (2011) Improving the profitability of willow crops—identifying opportunities with a crop budget model. Bioenergy Res 4:85–95

De La Torre Ugarte D, Ray D (2000) Biomass and bioenergy applications of the POLYSYS modeling framework. Biomass Bioenergy 18:291–308

Dykstra D (2004) RILSIM 2.0 user's guide: software for financial analysis of reduced-impact logging systems. Published as a contribution to the Asia-Pacific Forestry Commission, Blue Ox Forestry, Portland, OR, p 120

Greene WD, Lanford BL (1996) Logging cost analysis. Short course manual. Georgia Center for Continuing Education, Athens, GA, p 78

Greene WD, Lanford B (2001) Logging cost analysis. The University of Georgia, Georgia Center for continuing education and Daniel B. Warnell School of Forest Resources. Rural Development Center, Tifton, Georgia

Greene WD, Reisinger TW (1999) Harvesting and roads. Timber harvesting teaching packet. The University of Georgia, Athens, GA

Grushecky S, Wang J, McGill D (2007) Influence of site characteristics and costs of trucking and extraction on log residue utilization in southern West Virginia. Forest Prod J 57(7/8):63–67

Han HS, Harry WL, Johnson LR (2004) Economic feasibility of an integrated harvesting system for small-diameter trees in southwest Idaho. Forest Prod J 54(2):21–27

Hartsough B, Yomogida D, Stokes B (1996) Harvesting systems for short rotation woody crops. First conference of the short rotation woody crops working group, Paducah, KY, September 23–25, 1996

Hartsough B, Drews E, McNeel J, Durston T, Stokes B (1997) Comparison of mechanized systems for thinning ponderosa pine and mixed conifer stands. Forest Prod J 47(11/12):59–68

Hendricks GL (1986) A user's manual for the harvesting analyzer: a microcomputer-based program for loggers. Division of Land and Economic Resources, Tennessee Valley Authority, Norris, TN, p 72

Howard A (1989) A sequential approach to sampling design for time studies of cable yarding operations. Can J Forest Res 19:973–980

Howard A, Gasson R (1991) A system for computer-based design and implementation of time studies. Forest Prod J 41(7/8):53–55

Jackson B, Schroeder R, Ashton S, Taylor E (2010) Chip-ping woody biomass. http://www.extension.org/pages/Chipping_Woody_Biomass. Accessed 12 May 2010

Jensen K, Menard J, English B, Park W, Wilson B (2002) The wood transportation and resource analysis system (WTRANS): an analysis tool to assist wood residue producers and users. Forest Prod J 52(5):27–33

Jeuck J, Duncan D (2009) Economics of harvesting woodybiomass in North Carolina. http://www.ces.ncsu.edu/forestry/biomass/pubs/WB011.pdf. Accessed 19 May 2010

Johansson J, Liss J, Gullberg T, Björhedenc R (2006) Trans- port and handling of forest energy bundles—advantages and problems. Biomass Bioenergy 30(4):334–341. doi:10.1016/j.biombioe.2005.07.012

Kingsley E (2011) Understanding biomass supply and demand. Northeast wood energy short course. November 8, 2011. State College, PA. Pennsylvania State University, State College, PA (presentation). http://www.bioenergy.psu.edu/shortcourses/2011WoodEnergy/Kingsley-PSUBiomass-INRS_Nov_2011.pdf. Accessed 11 April 2013

Li Y (2005) Modeling operational forestry problems in central Appalachian hardwood forests. Dissertation, West Virginia University, Morgantown, WV, p 144

Li Y, Wang J, Miller G, McNeel J (2006) Production economics of harvesting small-diameter hardwood stands in central Appalachia. Forest Prod J 56(3):81–86

Matthews DM (1942) Cost control in the logging industry. McGraw-Hill Book Company, New York, NY, p 374

McDonald T (1999) Time study of harvesting equipment using GPS-derived positional data. In: Proceedings of forestry engineering for tomorrow. Edinburgh University, Edinburgh, Scotland, June 28–30, 1999

McDonald T, Rummer B (2000) Automating time study of feller-buncher. In: Proceedings of the 23rd annual meeting of council on forest engineering, Kelowna, Canada, September 11–14, 2000

Missouri Forest Products Association (2010) Woody biomass technology—demonstration project. http://MFPA%20woody%20biomass%20community%20demonstration%20report.pdf. Accessed 19 May 2010

Miyata ES (1980) Determining fixed and operating costs of logging equipment. US For. Serv. Gen. Tech. Rep. NC-55. p 16

Miyata ES, Steinlulb HM (1981) Logging system cost analysis: comparison of methods used. Res Pap NC-208, US For. Serv., North Cent. For. Exp. Stn., St. Paul, MN. p 15

Olsen E, Kellogg L (1983) Comparison of time-study techniques for evaluating logging production. Transact ASAE 26(6):1665–1668

Perlack RD, Wright LL, Turhollow AF, Graham RL, Stokes BJ, Erbach DC (2005) Biomass as feedstock for a bioenergy and bioproducts industry: the technical feasibility of a billion-ton annual supply. Oak Ridge National Laboratory, Oak Ridge, TN

Reisinger TW, Gallagher TV (2001) Evaluation and comparison of two tree-length harvesting systems operating on steep slopes in West Virginia. In: Wang J, Wolford M, McNeel J (eds) Proceedings of the 24th annual meeting of the council on forest engineering: Appalachian hardwoods: managing change, July 15–18, 2001, Council on Forest Engineering, Snowshoe, WV, pp 56–64

Rummer B, Len D, O'Brien O (2004) Forest residues bund-ling project—new technology for residue removal. USDA forest service, southern research station, Auburn, AL. http://www.fs.fed.us/woodybiomass/strategy/bundling/documents/bundler_report_final.pdf. Accessed 16 May 2009

Scott DA, Dean TJ (2006) Energy trade-offs between intensive biomass utilization, site productivity loss, and ameliorative treatments in loblolly pine plantations. Biomass Bioenergy 30(12):1001–1010. doi:10.1016/j.biombioe.2005.12.014

Sokhansanj S, Kumar A, Turhollow A (2006) Development and implementation of integrated biomass supply analysis and logistics model (IBSAL). Biomass Bioenergy 30(10):838–847

Spinelli R, Nati C, Magagnotti N (2007) Recovering logging residue: experiences from the Italian Eastern Alps. Croat J Forest Eng 28:1–9

Stenzel G, Walbridge TA, Pearce JK (1985) Logging and pulpwood production, 2nd edn. Wiley-Inter-Science Publisher, New York, p 358

Stokes BJ, Schilling A (1997) Improved harvesting systems for wet sites. Forest Ecol Manag 90:155–160

Stuart WB (1981) Harvesting analysis technique: a computer simulation system for timber harvesting. Forest Prod J 31(11):45–53

Tufts RA, Mills WL Jr (1982) Financial analysis of equipment replacement. Forest Prod J 32(10):45–52

Tufts RA, Lanford BL, Greene WD, Burrows JO (1985) Auburn harvesting analyzer. Compiler 3(2):14–15

Visser RJM, McDonaugh KD, Meller RD, McDonald TP (2004) Harvesting system simulation using a systems dynamic model. S J Appl Forest 28(2):91–99

Wang J (2007) Hardwood log bucking and loading efficiency in West Virginia. Forest Prod J 57(5):84–90

Wang J, Greene W (1999) An interactive simulation system for modeling stands, harvests, and machines. J Forest Eng 10(1):81–99

Wang J, LeDoux C (2003) Estimating and validating ground-based timber harvesting production through computer simulation. Forest Sci 49(1):64–76

Wang J, Greene WD, Stokes BJ (1998) Stand, harvest, and equipment interactions in simulated harvesting prescriptions. Forest Prod J 48(9):81–86

Wang J, McNeel J, Baumgras J (2003) A computer-based time study system for timber harvesting operations. Forest Prod J 53(3):47–53

Wang J, Long C, McNeel J (2004a) Production and cost analysis of a feller-buncher and grapple skidder in central Appalachian hardwood forests. Forest Prod J 54(12):159–167

Wang J, Long C, McNeel J, Baumgras J (2004b) Productivity and cost of manual felling and cable skidding in central Appalachian hardwood forests. Forest Prod J 54(12):45–51

Wang J, LeDoux C, Li Y (2005) Simulating cut-to-length harvesting operations in Appalachian hardwoods. Int J Forest Eng 16(2):11–27

Wu J, Wang J, McNeel J (2011) Economic modeling of woody biomass utilization for bioenergy and its application in central Appalachia. Can J Forest Res 41(1):165–179

Wu J, Wang J, Li Y, Spong B (2012) A web-based decision support system for analyzing timber harvesting costs and productivity. N J Appl Forest 29(3):141–149

Chapter 6
Residential Wood Heat in the Northeast

Corie Podschelne and Daniel Ciolkosz

6.1 Introduction

No discussion of wood heat in the Northern Forest would be complete without giving due consideration to the role of residential wood heat. Not only is it important historically, but wood for residential heat continues to be an important component of the region's cultural, ecological, and energetic status.

6.1.1 Preview of Chapter

This chapter presents the current status of residential wood heat in the Northeast USA, discussing three main topics:

- How wood heat is being used in the region
- What technologies are employed to provide wood heat to homes
- Prospects for future trends in wood heat

The premise behind this chapter is the belief that wood heat continues to be an excellent means for providing sustainable heat for many homes in the forested region of the Northeast USA, and that future developments will allow wood heat to be more efficient, effective, and clean.

C. Podschelne (✉)
Harman Home Heating, 352 Mountain House Road, Halifax, PA 17032, USA
e-mail: PodschelneC@hearthnhome.com

D. Ciolkosz
Department of Agricultural and Biological Engineering, The Pennsylvania State University, 249 Agricultural Engineering Building, University Park, PA 16802, USA

M. Jacobson and D. Ciolkosz (eds.), *Wood-Based Energy in the Northern Forests*,
DOI 10.1007/978-1-4614-9478-2_6,

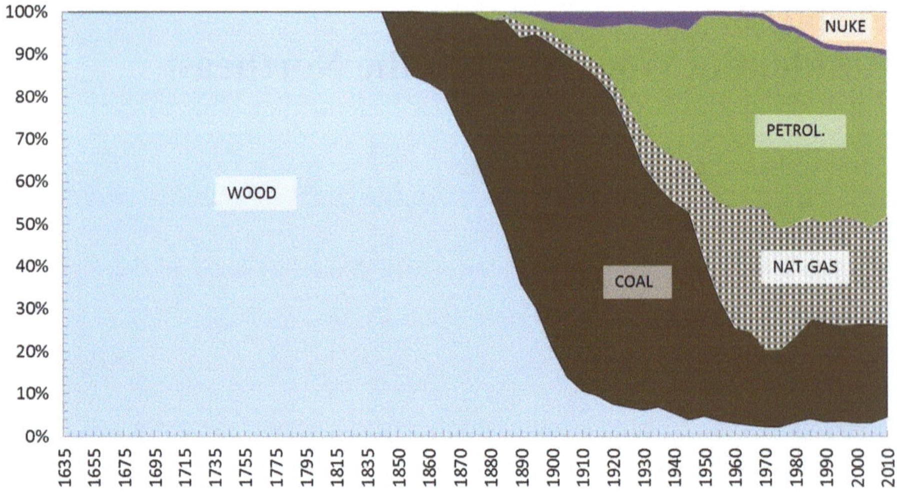

Fig. 6.1 Energy use of the American colonies, and later the United States, 1635–2010

6.1.2 Historical Aspects of Residential Wood Heat in the Region

Wood heat is the original home heating fuel for North America, and remains a popular means for keeping warm throughout the Northeast USA. Until the mid1800s, wood was essentially the only fuel used, after which time Coal, then Natural Gas, Petroleum, Hydropower, and Nuclear began to contribute noticeably to America's energy supply (see Fig. 6.1). It is interesting to note that while the current fraction of the nation's energy that is provided by wood is small, the total quantity of wood energy (i.e., in Exajoules) is at record high levels—even when compared to the 1800s when forests in the Northeast were heavily over-harvested and depleted to meet industrial heat and lumber demand. Since that time, forests have largely recovered in the region, and currently see their greatest pressure not from harvesting but from land use pressures due to the large and growing population in the region (see Chap. 1 for a thorough discussion of the state of the forest in the region).

While open fires and fireplaces were originally the means of providing heat to homes, their low efficiency led to developments such as the celebrated "Franklin Stove" of colonial America, which was the precursor of the modern wood stove. Developed by Benjamin Franklin in 1741, the Franklin Stove was essentially a cast iron fireplace insert that used a heat exchanger to transfer more of the fire's heat to the occupied space. With a few minor improvements from others, the Franklin Stove soon became a common feature in homes throughout colonial America.

6.2 Use of Wood Heat in the Northeast

6.2.1 *Spatial Variation*

According to recent government estimates, about half a million homes in the Northeast USA use wood as their primary heating fuel. This amounts to about 21 % of the national total. Wood heat is used in all states of the region, but the distribution of usage varies greatly. The highest percentage of homes that use wood heat is in the highly forested states of Vermont, Maine, New Hampshire, and West Virginia. However, when the populations of the region's states are factored in, we find that the largest number of homes using wood is in New York and Pennsylvania, together accounting for nearly half of the region's wood heated homes. Wood heat use is generally more prevalent in rural areas, with rural regions 15 times more likely to use wood heat than urban dwellings (Anon 2011) (Fig. 6.2).

Wood is typically either cut by the user, or is delivered by small businesses that specialize in cordwood cutting and delivery. Wood pellets are an exception, in that they are typically sold by the bag or pallet at retail stores. Wood pellets remain a small percentage of the total amount of wood heat in the region, although their use has increased noticeably since the turn of the century (Spelter and Toth 2009).

6.2.2 *Trends Over Time*

While wood was originally the exclusive source of home heat in the region, its use dropped during the twentieth century, reaching a low point in the late 1960s and

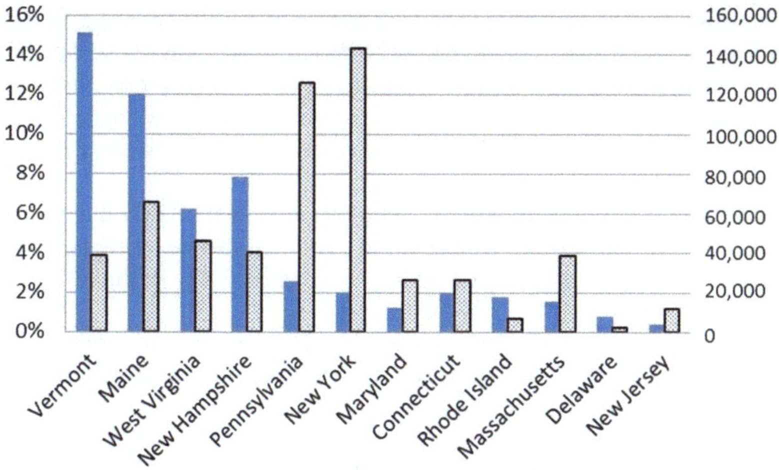

Fig. 6.2 Percent (*dark bars*) and number of homes (*light bars*) using wood heat in the Northeast

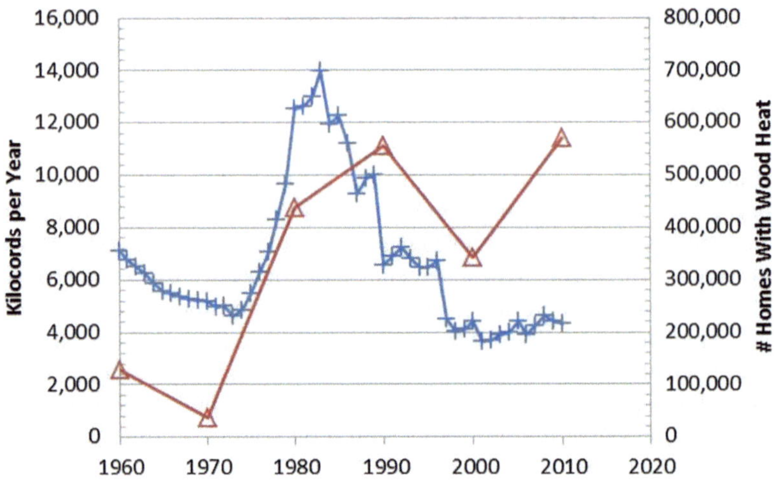

Fig. 6.3 Amount of wood (*plus*) and number of homes using wood (*triangle*) in the Northeast, 1960–2010

1970s. A surge in popularity more than doubled wood heat use in the late 1970s and early 1980s (corresponding to the rise in petroleum prices during that period), but has steadily decreased since. While petroleum prices have again risen in real dollars since about 2004, a corresponding rise in wood heat use has not yet occurred as of 2010, although the number of homes using wood heat has risen in the region. This suggests two possibilities: either rising oil prices have become a less important driver of wood heat use in the region; or increased efficiency of wood burning equipment and/or improved energy efficiency of homes is reducing the amount of wood used per home. The second factor appears to be the more likely explanation, and it has several interesting implications—first, that our wood resources can now heat many more homes than only a few years ago, and second, that emissions resulting from of incomplete combustion are less of a factor than in the past (Fig. 6.3).

When wood usage is compared to other regions in the USA during the past half century, the Northeast, Northcentral, and Southwest regions have both increased their relative share of wood heat use, while usage in the Southeast has dropped.

Currently, wood use for residential heat in the Northeast USA is overwhelmingly in the form of cordwood (logs). However, while wood pellet use as a fraction of total wood use remains small, the market continues to grow, as customers find that the ease of use and high efficiency of the combustion equipment outweighs the higher cost for wood pellets, relative to cordwood. Wood chips are a common fuel for Industrial and Commercial Scale boilers in the region, but they are practically non-existent at the residential scale.

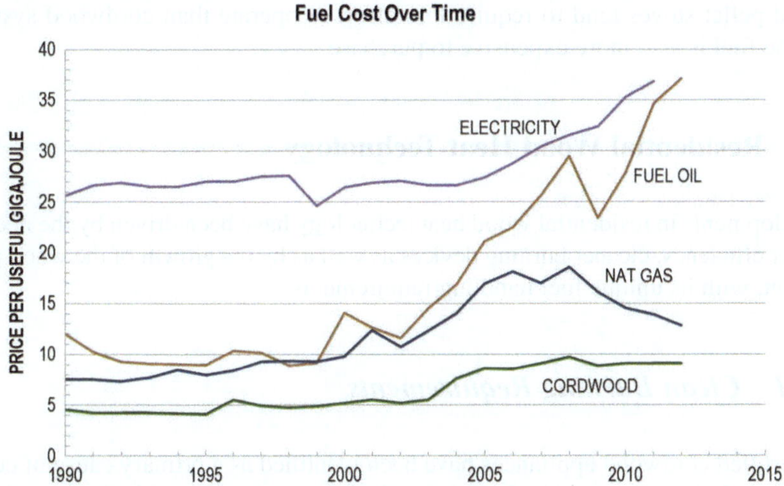

Fig. 6.4 Cost of heat from various fuels in the Northeast (Ciolkosz 2013)

6.2.3 Costs Relative to Other Heating Fuels

Wood has long been, and continues to be a cost-effective source of heat for homes in the region. While self-harvested firewood can be extremely inexpensive, it is only practical for a small segment of the community. Purchased firewood, delivered cut and split, is the more common option, and it has compared favorably to other fuels in terms of usable heat per dollar cost. Wood is commonly sold by the cord, a traditional unit of measure amounting to 3.6 m^3 of "racked and well stowed" firewood. While wood density, moisture content, and tightness of stacking can vary considerably (a firewood seller's idea of well stacked wood is rarely the same as a buyer's), we will use the federal government's working definition of a "standard cord" as having 21 gigajoules of energy (HHV) and 1.06 oven dry tonnes of material per cord.

When compared to other home heating fuels, wood is a very good value, and has been consistently so for many years. Analysis of advertized prices of cut, split, and delivered firewood in Central Pennsylvania (roughly representative of the region), relative to open market prices for other heating fuels, shows the significantly lower price that is paid for wood fuel. The key to valid comparisons is to determine the cost per amount of useful heat energy provided (dollars per useful gigajoule)—this accounts for fuel cost, the amount of heat in the fuel, and the typical efficiency of the heating equipment in delivering that heat to the home (Fig. 6.4).

In addition to its lower cost, wood fuel prices also show less volatility over time, which is a further benefit of the fuel. However, the lower cost of wood fuel must be weighed against the work that must be done stacking and loading wood, tending the fire, and cleaning ashes from the stove. Most fossil fuel systems, by comparison, are fully automated and require no effort on the part of the homeowner.

Wood pellet stoves tend to require less effort to operate than cordwood systems, but the fuel is also more expensive to purchase.

6.3 Residential Wood Heat Technology

Developments in residential wood heat technology have been driven by the need for higher efficiency, cleaner burning devices as well as by the growth of the wood pellet market, with its unique fuel handling requirements.

6.3.1 Clean Burning Requirements

Uncertified cordwood appliances have been identified as a primary cause of certain types of air pollution, particularly because they can be operated at extremely low burn rates and lack technology to reburn or recover the unused fuel contained in smoke. Therefore, in 1988, all new residential wood fueled appliances (with certain exemptions) offered for sale have been required to meet EPA regulations for smoke output or more accurately, particulate matter. In order to be sold, all wood stoves must be tested by an accredited third party laboratory and must not exceed the published EPA limits for particulate emissions. In the early stages of EPA regulation of wood stoves, many manufacturers struggled to design stoves which could comply with the regulations. The industry has experienced significant evolution since the promulgation of the first EPA regulations, and a variety of unit types and clean burning technologies are now available (Fig. 6.5).

Fig. 6.5 Modern clean burning woodstove

Certain units, deemed "non-affected facilities" by the EPA, are not subject to the same clean burning regulations as other appliances. Included in the list of non-affected facilities are wood stoves with an air to fuel ratio greater than 35:1, open masonry fireplaces, boilers, furnaces, and cookstoves. Although not required, many manufacturers offer clean burning versions of these non-affected facilities and the technologies used for emissions reduction in these units will be discussed later in this chapter.

Clean burning wood stoves also have significantly increased thermal and combustion efficiency, which results in significantly decreased wood consumption. Modern freestanding wood stoves, for instance, typically operate at 60–80 % overall efficiency, i.e., at least 60–80 % of the usable energy in the fuel wood is converted to heat in the room. Compare this with typical overall efficiencies of 40–60 % or less for older, uncertified stoves, and it is easy to understand why modern wood stove users typically report a 30 % reduction in wood consumed.

6.3.2 Clean Burning Technologies

6.3.2.1 Catalytic

Similar to the catalytic converter in an automobile, the wood stove catalytic combustor is a ceramic or stainless steel honeycomb which is coated with a precious metal, normally palladium or platinum. As the name implies, the catalytic material lowers the re-ignition temperature of the unburned smoke in wood stove exhaust from 1,000 °F to about 500 °F. This catalytic reaction, combined with a supply of oxygen, allows the catalytic stove to achieve highly efficient, clean burns. Catalytic stoves require a bypass damper, e.g., a device which allows the wood smoke to bypass the catalytic element while the stove is being started or reloaded. The catalytic stove excels at long, slow burns at moderate heat output levels, but normally with very little flame activity in the firebox. This technology is utilized in freestanding wood stoves, fireplace inserts, furnaces and boilers.

6.3.2.2 Non-catalytic

Tube Secondary Combustion

The most popular of all wood stove clean burning technology, the tube secondary combustion stove uses insulation, firebox flow dynamics and some type of air introduction system to distribute preheated secondary combustion air to the area above the burning fire. This secondary combustion air causes the unburned smoke of the wood stove to reignite brilliantly in a light show of floating, rolling flames. The high firebox temperatures required for this system to work also helps keep the viewing glass clean. The simplest to use of all the clean burning systems, tube secondary stoves excel at providing moderate burn times with extremely attractive flame patterns in the firebox. This technology is most commonly used in freestanding wood stoves and fireplace inserts, but also in some furnaces.

Gasification

Gasification technology is most commonly used in wood boilers, but also seen in a slightly different configuration in some freestanding stoves and inserts. There are different types of gasification configurations, but downdraft and updraft are the most commonly used in wood appliances. Although the construction of the units differs depending on the type of gasification being used, the concept is the same: A highly insulated firebox uses combustion air delivered at specific locations to gasify the combustible compounds in cordwood. Additional preheated combustion air is then introduced to the wood gas at the firebox top (in an updraft gasifier) or at the nozzle (in a downdraft gasifier) which ignites the hydrocarbon rich exhaust, resulting in very clean and efficient combustion. The heavy insulation of typical gasification units and the controlled combustion means these units excel at long burns, but similar to catalytic units, gasification units typically have very little flame activity in the firebox, because the majority of the combustion takes place inside the gasification chamber. Just like a catalytic unit, most gasification units require a bypass mechanism which disengages the gasification system while starting a fire or reloading.

6.3.3 Types of Units

6.3.3.1 Freestanding Wood Stoves

The most common type of wood heater, the freestanding wood stove is essentially a firebox which is suspended by either legs or a pedestal. Freestanding wood stoves are flexible in installation location, and many stoves feature heat shielding which allows close installation to combustible walls. Most freestanding stoves have a glass window which allows a view of the burning fire. The freestanding wood stove can be constructed of a variety of materials which include steel, cast iron, soapstone or some combination of the three. Wood stoves are designed to be connected to Class A prefabricated chimneys or existing masonry chimneys. Freestanding wood stoves are available in a variety of sizes, which allow them to heat a single room for a few hours or an entire home 24 h a day. Features such as top or side loading, thermostatic air controls, blowers and different clean burning technologies differentiate brands and models (Fig. 6.6).

Fig. 6.6 Freestanding wood stove

6.3.3.2 Fireplace Inserts

Traditional open fireplaces are extremely inefficient in regards to both clean burning and heating efficiency. Open fireplaces are typically negatively efficient, i.e. they actually remove heat from the home while burning due to the large and uncontrollable amount of room air that is drawn up the chimney. The fireplace insert was designed to combat both of these issues and fireplace inserts are very popular in the Northeast USA where a large majority of homes have some type of open fireplace.

A fireplace insert is essentially a freestanding stove that typically has special convection channels built around the firebox and is made to slide into an existing masonry fireplace and use the existing fireplace chimney. Fireplace inserts are reliant on electric convection fans to push heated air into the room because the majority of the firebox normally sits inside the existing fireplace. Fireplace inserts transform inefficient and dirty burning fireplaces into heating powerhouses, and models are available that are sufficiently large to heat an entire home. Both freestanding units and fireplace inserts are often available with an optional spark screen which allows open door burning, an important consideration for some open fireplace owners who enjoy the sight and sound of an open fire (Fig. 6.7).

6.3.3.3 Wood Central Heat

Due to the status of the wood boiler as the previously mentioned non-affected facility, clean burning boilers are not as prevalent as clean burning freestanding stoves and inserts.

Fig. 6.7 Fireplace insert

The outdoor version of the uncertified wood boiler has been an important topic in the wood heating industry and has caused the EPA to review the exemption for boilers. Unfortunately, these large, uncertified outdoor wood appliances have the capability to operate at extremely low burn rates, often becoming a smoking and unwelcome nuisance in residential areas. An Internet search will turn up many municipalities that have instituted bans on outdoor wood boilers.

Luckily, clean burning boilers are available and often feature advanced control and burn systems which make burning wood easier and more convenient. These boilers feature exhaust gas sensing and computer control systems which are constantly adjusting burn parameters for a highly efficient, clean burn. For consumers wishing to retain some of the convenience of fossil fuel central heat, modern, electronically controlled clean burning wood boilers are the closest cordwood burning alternative. A close second, in terms of convenience, would be the EPA-certified wood furnace, particularly when coupled with a fossil fuel or electric backup heat source, which is often available on these units. Cordwood furnaces, like their pellet burning counterparts, are often very expensive, but grants and tax credits are usually available at the state or federal level to help offset the cost of these units (Fig. 6.8).

6.3.3.4 Pellet Units

Pellet units have become an increasingly popular wood based heating choice in the Northeast USA. Pellet units burn pelletized sawdust, which is compressed using high-pressure extrusion and bonded solely by the lignin content of the wood. Pellet fuel is normally purchased in 40 lb bags and is typically considered less messy and easier to store than cordwood fuel. The packing or storage density of pellet fuel is also significantly higher than cordwood 577 kg/m^3 versus 289 kg/m^3, which allows

Fig. 6.8 Clean burning updraft wood gasification boiler (photo courtesy of Greenwood Boiler, LLC)

Fig. 6.9 Wood pellet fuel (photo courtesy of Greene Team pellets)

the storage of more pellet fuel in a given amount of space (Ciolkosz 2010). Pellet fuel also generates less ash, with premium pellet fuel required to be less than 1.0 % ash, versus cordwood fuel which can be as high as 1.82 % ash (Misra et al 1993; PFI, 2010) (Fig. 6.9).

Pellet units, which can be freestanding stoves, inserts or central heat furnaces or boilers, are inherently clean burning. This is due to two primary factors: the extremely low moisture content of the fuel and the nature in which it is burned, which is typically in a small, crucible-like burn pot. Pellet stoves feature an electronic feed system, convection blowers, automatic ignition, and microprocessor controls, which have allowed the pellet stove to approach the convenience of a fossil

Fig. 6.10 Freestanding pellet stove

fuel heating system. The pellet stove user fills a fuel storage hopper on the unit with pellet fuel and the unit automatically feeds and ignites pellet fuel, often using a thermostat or room sensor to maintain a specific room temperature (Fig. 6.10). In the case of pellet central heat, both pellet boilers and furnaces are available which, with their large hoppers and ash pans, are the closest rival to fossil fuel central heat in terms of automation and control capabilities.

The low exhaust temperature of pellet units allows the use of a much lower cost version of venting pipe, known as Type P or Type L pipe and the venting termination of a pellet chimney system does not have to penetrate the roof of a house, like a wood stove chimney. This allows its appearance on the outside of the house to look more like a clothes dryer vent than a heating exhaust termination, further increasing the appeal to many homeowners. The advantages of the pellet stove also include higher overall efficiencies compared to freestanding wood stoves, which pellet stoves accomplish using aggressive heat exchangers in conjunction with their inherently high combustion efficiency, which is normally greater than 95 %.

6.3.4 Prospects for the Future of Wood Heat

As air quality in pollution-prone air sheds continues to be an issue, regulations regarding clean burning requirements for wood fired units will continue to tighten. The current emission standards are being studied and recommendations for significantly lowered emission outputs are being driven at the state and federal level. Potentially, this could mean future residential wood heat use will likely depend heavily upon wood burning regulations, from the federal to the local level. It is not uncommon for localities to introduce tiered burn bans based on air quality, but

EPA-certified wood and pellet stoves are frequently allowed during all but the worst air quality burn bans. As the clean burning requirements become stricter, the technology required for many units to comply has the potential to become cost prohibitive at the consumer level. Several manufacturers have already begun rolling out hybrid units which combine two types of the clean burning technology, to help achieve ultraclean combustion and comply with upcoming federal regulations.

Currently, the availability and cost of cordwood fuel are the driving force behind the popularity of wood fueled appliances in this region. Cordwood is one of the cheapest alternatives to fossil fuel, with wood pellet fuel a close second. The high cost savings realized with these two fuels allows consumers switching from fossil fuel heat to have a short return time on their investment, particularly in the case of lower priced, EPA-certified wood units. Fuel availability has been a volatile issue in the pellet market, but manufacturers appear to be developing a better grasp of the ebb and flow of the wood pellet market and fuel shortages have been minimized in recent years. Cordwood availability varies by locale, but cordwood availability, much like wood pellet availability, can have a significant impact on the future of wood heat; additional information on cordwood fuel availability is discussed in Chap. 1.

Lastly, the convenience of fossil fuel and alternative heat sources and their cost relative to cordwood will certainly have an effect on the prevalence of wood heat in the future. As society moves towards faster and easier means of doing almost everything, wood heat is decidedly inconvenient. The whole process of felling trees, bucking logs, splitting, stacking, and seasoning cordwood is certainly not achieved with an application on a smartphone. Fossil fuel alternatives including geothermal, solar, and even air source heat pumps offer varying, but often longer return on investment periods. However, as the technology for these heat sources becomes less expensive, the convenience of these alternatives may well be worth the initial investment to those which would have normally considered wood heat. Conversely, wood combustion engineers are faced with the challenge of creating wood heat systems that have better performance characteristics and lower maintenance needs, so as to better compete with these alternatives and meet the needs of society with a renewable, sustainable, clean source for residential heat.

References

Anonymous (2011) The 2006–2010 American community survey data file. Version 2. December 2011. United States Census Bureau. http://www2.census.gov/acs2010_5yr/summaryfile/UserTools/. Accessed Mar 2013

Anonymous (2012a) State energy consumption estimates 1960 through 2010. United States Energy Information Administration. DOE/EIA-0214(2010). June 2012

Anonymous (2012b) Annual energy review 2011. United State Energy Information Administration. Report # DOE/EIA-0384(2011). September 2012

Anonymous (2013) Historical census of housing tables. United States Census Bureau. Online document: http://www.census.gov/hhes/www/housing/census/historic/fuels.html. Accessed Mar 2013

Ciolkosz, D (2010) De-mystifying wood as a renewable fuel.http://www.bioenergy.psu.edu/shortcourses/2010_nov_chp/04_demystifying.pdf

Ciolkosz D (2013) Wood heat for the home: does it pay off? Penn state extension renewable and alternative energy fact sheet series. University Park, PA

HPBA (2008) Clearing the smoke: the Libby woodstove changeout program. http://www.woodstovechangeout.org/fileadmin/PDF/Libby_Report-Final.pdf

http://www.bioenergy.psu.edu/shortcourses/2010_nov_chp/04_demystifying.pdf

Misra MK, Ragland KW, Baker AJ (1993) Wood ash composition as a function of furnace temperature. Biomass Bioenergy. 4(2)103–166

PFI (2010). Pellet Fuels Institute Standard Specification for Residential/Commercial Densified Fuel. Oct 25, 2010. Pellet Fuels Institute. Arlington, VA

Spelter H, Toth D (2009) North America's wood pellet sector. USDA forest service research paper FPL-RP-656

Chapter 7
Commercial Combustion and CHP Systems

Daniel Ciolkosz and Jim Babcock

7.1 Introduction/Overview

Wood heat for individual homes (i.e., wood stoves and pellet stoves) is widely recognized and understood in the Northeast USA. Commercial-scale wood heat and CHP (combined heat and power), however, is not nearly as well known or as widely used. Nonetheless, it is a significant opportunity that continues to experience growth in the region. Commercial-scale wood heat or CHP systems are designed to burn wood, producing heat or heat + electrical power for medium sized facilities, such as schools, office buildings, hospitals, and manufacturing plants.

Currently, commercial scale wood combustion systems occupy a favorable niche in the forest bioenergy picture for the Northeast USA, due to the following factors:

- the plentiful supply of low value wood from forests in the region,
- the relatively close proximity of the region's population to the forest resource,
- the rising and variable costs of fossil fuels, and
- the high heating load of many facilities in the region.

This is a compelling argument in favor of using wood to provide heat or heat + power to small to medium sized commercial and institutional facilities. Originally, these systems were almost exclusively used by the wood products industry. However, starting in the 1990s, schools, hospitals, and other commercial facilities in the region have begun to switch to wood heat, primarily for economic reasons,

D. Ciolkosz (✉)
Department of Agricultural and Biological Engineering, The Pennsylvania State University, 249 Agricultural Engineering Building, University Park, PA 16802, USA
e-mail: dec109@psu.edu

J. Babcock
Advanced Recycling Equipment, Inc, 850 Washington Rd., St Marys, PA 15857, USA
e-mail: jbareinc@windstream.net

M. Jacobson and D. Ciolkosz (eds.), *Wood-Based Energy in the Northern Forests*,
DOI 10.1007/978-1-4614-9478-2_7,

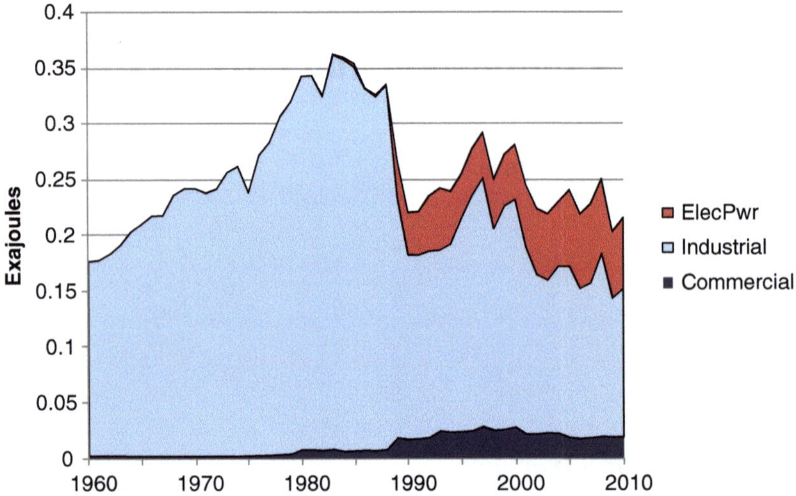

Fig. 7.1 Commercial, industrial, and power sector wood use in the Northeast USA, 1960–2010

but also for the perceived value of utilizing locally produced, renewable fuel. This has been most notable in Vermont and Pennsylvania, two states with active "Fuels for Schools" programs. An exhaustive survey of wood heat systems in the region has not been completed, but the number of systems in use is generally less than 100 per state. US Census statistics indicate that, while recent installations of commercial-scale wood combustion systems have been seen at several locations in the region, overall use of wood (by the commercial sector and the industrial sector) is, if anything, declining (Fig. 7.1).

Most commercial scale wood combustion systems in the region are "heat only" plants, but there are a few examples of combined heat and power production as well.

Recent growth notwithstanding, adoption of commercial scale wood heat is by no means widespread, and several technical challenges must be addressed if sustainable wood heat and CHP are to expand their role in the region's energy economy. As these issues are addressed by continued research and product development, we can expect sustainable growth of wood heat and CHP through the twenty-first century.

7.2 Typical System Configurations

Commercial wood combustion systems are comprised of several distinct components, all which are needed for the entire system to operate properly. Typically, a commercial wood combustion system will include (1) a fuel storage and handling system, (2) a combustor, (3) a heat exchanger, (4) pollution control devices, (5) an ash collection system, and (6) an exhaust stack. A building is usually also needed to

Fig. 7.2 Typical commercial-scale biomass combustion system, with combustor and boiler on the right, multiclone and ash collection on the left (Photo Courtesy Advanced Recycling Equipment Inc.)

house these components, as well as a connection to the facility's heat distribution system. The system can be fully integrated into the building's architecture, or can be designed as a separate structure, depending on the needs of the owner and constraints of the site. Each subsystem must be appropriately designed and operated in order for the system to function well (Fig. 7.2).

7.2.1 Fuel Storage and Handling

Fuel storage and handling is a critical and unique component of commercial wood combustion systems. Their design depends primarily on the type of fuel used—green wood chips (~40 % mc, wet basis) are the most common fuel, and are stored in either aboveground silos or belowground pits. Augers, walking floors, and/or belts are used to move fuel from the storage area to the combustor. These storage areas are typically designed to hold sufficient fuel for 3–5 days of operation—fuel suppliers are contractually obligated to maintain a steady supply of fuel. Because of the relatively short storage time of the fuel, self-heating of the fuel is usually not an issue. However, larger facilities sometimes opt for on-site outdoor storage piles, with skid loaders or front end loaders used to move the fuel to a "day bin" from which the fuel is fed by augers into the combustor. If uniformity of wood chip size is an issue, it may be prudent to include a "shaker/screener table" to remove oversized pieces. Magnetics should also be incorporated into the fuel transfer system if tramp metal could be an issue. Occasionally, a grinder is included in the system as well. Fuel delivery trailers are occasionally used as the storage system, but it is often more economical to construct a dedicated storage facility on site. While indoor storage pits are common for wood chip storage, aboveground silos can also be used, but care must be taken to ensure flowability of the chips from the silo, and to avoid freezing of large chunks of fuel, which can jam the fuel handling system.

If wood pellets are the fuel of choice, aboveground storage silos are typically used for fuel storage, and flexible augers are used to move the fuel to the combustor. Handling systems for wood pellet fuel are generally less expensive than those for wood chips.

Occasionally, sawdust is the fuel of choice, in which case storage and handling is similar to that for wood chips, due to its similarity to wood chips in bulk density and heat value. In all cases, care must be taken to control airborne dust as it poses both an inhalation and a fire hazard. Typically, a fuel storage and handling system designed for wood chips can be utilized for wood pellets as well, but not vice versa. This should be taken into consideration if the installed combustion system has the ability to utilize multiple fuel types.

Some facilities choose to collect low cost "roadside tree trimmings" and other waste wood for fuel. This wood is typically more prone to bark and dirt, and often has less size uniformity. As a result, additional regrinding and screening equipment may be needed.

7.2.2 Combustor Design

The combustor is the compartment in which fuel and air are "reacted," releasing heat. The overall reaction equation using an approximate formula for wood of $C_6H_{10}O_5$) is deceptively simple:

$$C_6H_{10}O_5 + 6O_2 \rightarrow 6CO_2 + 5H_2O$$

The actual process of combustion is actually very complex, involving a series of reactions and the production of intermediate products. At a very rough scale, three main processes dominate the combustion of wood: drying (water evaporates from the wood cells), flaming combustion (portions of the wood devolatilize into combustible vapors that react with oxygen, forming flames), and char combustion (portions of the wood remain in a solid form, reacting with oxygen to give off heat. Water in the wood fuel and nitrogen in the combustion air are also present in the combustor, as well as small amounts of minerals in the fuel, which are converted during combustion to water vapor and mineral oxide ash. Heat output from the combustion of bark free wood is approximately 20 GJ/ton (higher heat value), with less than 1 % ash content (on a mass basis) (Fig. 7.3).

Combustors are usually lined with ceramic refractory material, which serves to contain the heat and combustion gases, as well as to protect the steel superstructure

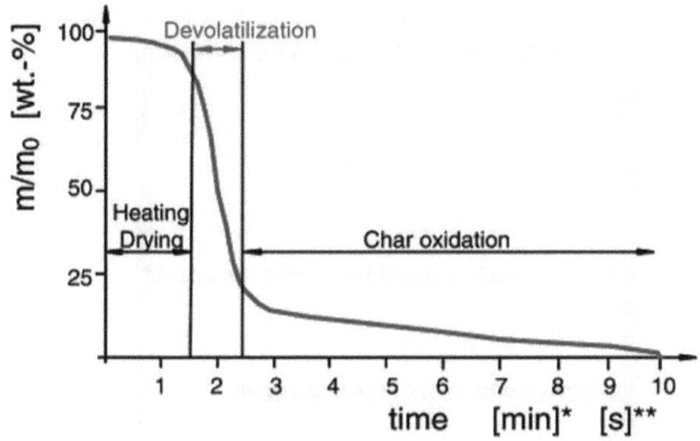

Fig. 7.3 Typical mass loss over time for wood combustion, showing the three processes dominating at different points (from Nussbaumer 2003)

of the combustor. Air is fed into the combustor in precise amounts at specific locations, so as to provide the needed air for the reaction.

Combustor design can be split into two main categories: direct combustors and indirect or "gasifier" combustors.

7.2.2.1 Direct Combustors

Direct combustors are devices in which the fuel is heated, dried and combusted all in one compartment. The residence time of fuel within the combustor depends on the expected fuel size, but is typically several minutes for wood chips. The residence time for air within the combustor is generally about 1.2–1.5 s, to allow for complete combustion of volatilized gases. Air is fed at a precise rate into the combustor to allow for complete combustion, with an excess air ratio (fraction of air relative to the stoichiometric amount) usually on the order of 1.1–1.2. Air is usually fed into the combustor at several locations, which helps to ensure complete combustion. An oxygen sensor in the stack is sometimes used to control airflow into the combustor. Direct combustors are the most common type of combustor available on the market today, and there are several types of designs, categorized by the manner in which fuel is fed into and through the combustor (Nussbaumer 2003). Some of the more common types include the following:

Fixed grate: Fixed grate combustors place the fuel on a static grate (often sloped), with combustion air provided both below and above the grate. Ash falls through or slides down the grate as the fuel burns, where it is collected below and removed (often manually). Fixed grate combustors are one of the older configurations in use today. These systems are well suited for variable-size fuel, but are not able to control the application of combustion air as precisely as some other configurations.

Moving grate: In a moving grate combustor, fuel is moved into and through the combustor on a metal grate that slowly slides or shakes the fuel along through the

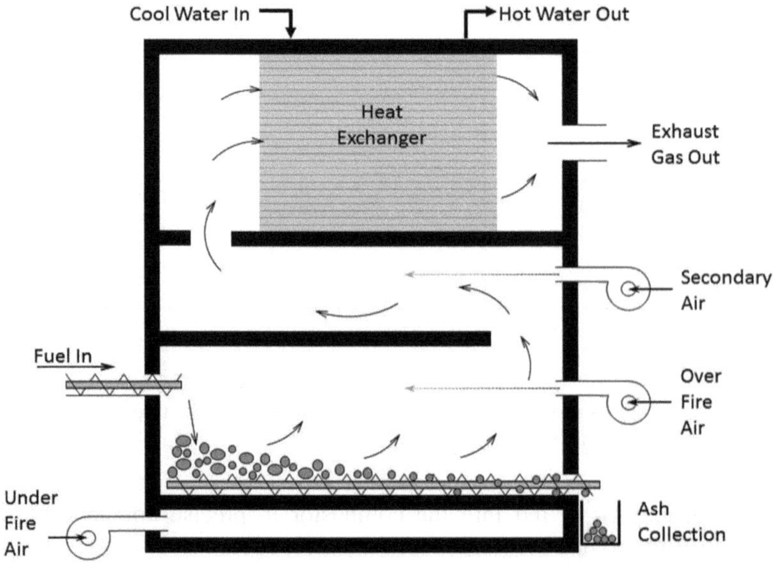

Fig. 7.4 Diagram of moving grate combustor with integrated fuel and ash handling system

combustor as it burns. This configuration is more successful than a fixed grate for fine-tuning the delivery of combustion air, but requires a fuel of regular size and composition. Auger floor systems are a type of moving grate, except that instead of carrying the fuel on a suspended grate, the fuel is pushed along the floor of the combustor using high temperature steel augers that rotate slowly as the fuel is fed into the combustor. This approach has grown in popularity recently for small to medium sized commercial applications, partly because of the manner in which they integrate fuel feed and ash handling into a single automated system (fuel is pushed in at one end, and ash is pushed out the other end) (Fig. 7.4).

Suspension/entrained combustion: Suspended fuel systems use very small particle size fuel such as sawdust. They blow the fuel into the combustor, where it combusts in a "ball of fire." One common style of suspension combustor is designed to swirl the fuel around in a large vortex in the cylindrical combustion chamber—they are often seen in use in the wood products industry.

Fluidized bed: A fluidized bed combustor burns fuel by adding it to a continually stirred bin of heated sand-like material. This arrangement allows for extremely even heating of the fuel and high efficiency of combustion. It also gives the combustor a great deal of flexibility as to which fuels to use. However, the system is expensive to purchase and operate.

Other configurations, such as the spreader stoker and the under feed stoker, are variations of the above. Each feed configuration has advantages and disadvantages, although the efficient implementation of a given design is probably more important than the type of feed configuration.

7.2.2.2 Indirect/Gasifier Combustors

Gasifier combustors are a less common type of combustor that operates using a distinct two step process. In the first step, wood fuel is heated in an oxygen-free compartment. At this point the molecular bonds in the wood will thermally activate and break down into a combustible gas called "syngas" or "producer gas." The composition of syngas can vary depending on the fuel and the equipment, but is primarily methane, carbon monoxide, carbon dioxide, and water vapor. In the second step of the process, the combustible gas is mixed with air and burnt under controlled conditions to produce heat, much in the same way that natural gas or propane is used as a combustion fuel. Gasifier combustors are more expensive and complex than direct combustion units, but they offer the potential for precise control of the combustion process, reduced slagging and fouling problems, and shorter response time to variable loads. In addition, since ash is created and removed prior to combustion, particulate emissions from gasifier combustors tend to be much lower than from direct combustors, if other emission controls are not used.

7.2.3 Heat Exchanger

The heat exchanger (often called a "boiler") is used to transfer heat from the hot combustion gases to the heating fluid (which is almost always water). The most common heat exchanger is of a shell and tube style, and can be designed either to produce hot water or steam, depending on facility needs. Ash buildup in the heat exchanger is a maintenance issue, and many manufacturers use air-operated soot blowers to automate the process of cleaning the heat exchanger. Occasionally, ash may have a tendency to fuse together on surfaces in the heat exchanger. This process, known as "fouling," is especially troublesome for system maintenance, but is rarely a problem when using wood as a fuel, provided that dirt in the fuel is kept to a minimum.

7.2.4 Pollution Control

In addition to the normal CO_2 and water vapor that are byproducts of combustion, Emissions from wood combustion systems can include uncombusted carbon and hydrocarbons, nitrogen oxides (NOx), and fine particulates (ash). Emissions are controlled by (1) maximizing combustion efficiency to eliminate the emission of carbon and hydrocarbons, (2) controlling combustion conditions to reduce emissions, and (3) using "filtering" devices to remove any remaining material from the stack gas before it is exhausted to the atmosphere.

The most commonly discussed controllable pollutant from wood combustion is particulate matter (ash). Reduction of particulate emissions is usually achieved through use of a "multi-clone" cyclonic separator. If additional reductions in particulate levels are appropriate, "bag houses" of filters, and/or electrostatic precipitators (ESPs) may be added as well. The northeast is currently seeing a trend away from bag houses towards ESPs, due to their lower maintenance and operating costs. These devices are typically installed downstream of the heat exchanger, and prior to the exhaust stack.

Nitrogen oxide (NO*x*) emissions result from the reaction of nitrogen (from the fuel or air) and oxygen in the high temperature environment of the combustor (Jenkins et al. 2011). NO*x* emissions can be controlled somewhat by modifying the rate of fuel feed, air supply to the combustor, and controlling combustion temperature (Nussbaumer 2003). NO*x* emissions are not commonly controlled in commercial wood combustion systems, because emission levels are usually well within regulated limits. Carbon monoxide is also a product of the combustion reaction but can be controlled through effective distribution of the combustion air and a controlled combustion temperature. Lower combustion temperatures also help to reduce the emission of fine particulates, by reducing the evaporation of mineral oxides that can recondense in the stack as extremely fine particles. Wood fuel is normally free of sulfur, so emission of sulfur oxides (SO*x*) is not a concern. Water vapor emissions are also not usually a matter of concern.

While interest has grown in the control of CO_2 emissions, an effective method for extracting CO_2 from the exhaust stack remains to be seen. Actual carbon dioxide and greenhouse gas emissions from biomass combustion are a surprisingly complex issue that remains a matter of debate in the academic community (i.e., Whitmann and Lehmann 2011; Johnson 2009)—a clear consensus has yet to emerge. At any rate, if the harvested wood that is used for fuel is being replaced by new forest growth, the use of wood for fuel can be thought of as being sustainable and probably very close to carbon neutral in the long term. See Chap. 3 for a more thorough discussion of this issue.

7.2.5 Electrical Power Production

Electrical power production from wood is most commonly achieved using steam from the heat exchanger coupled to steam turbines that generate rotational energy that in turn operates electrical generators. Practical steam turbines are available in sizes down to about 100 kW (output shaft power), and operate at pressures of 700 kPa (~100 psi) and above. The turbine and attached generator can be built in a number of configurations, the most common for wood systems being synchronous and asynchronous turbine generators. Synchronous generators (like synchronous motors) use magnets and windings to generate alternating current electrical power with a frequency precisely proportional to the rotating speed of the device. Asynchronous generators (also called "induction" generators) do not use magnets, but instead apply an excitation current to the windings to "induce" an electromagnetic field in the generator, causing it to rotate. Once this happens, additional rotational force from a steam turbine can be applied to the generator shaft, causing it to rotate slightly faster than the induced field. This results in the generation of electrical power.

Another approach to power production is to utilize a "hot gas turbine"—essentially a jet engine that produces rotational power for generating electricity instead of producing thrust. This requires that the wood fuel be converted to a combustible gas—via gasification—before being fed into the turbine. Lastly, it is also possible to utilize a reciprocating engine generator (a diesel engine, essentially) that runs on gasified wood. These two approaches have yet to achieve widespread acceptance, and steam-based generation remains dominant.

7.2.6 Ash Collection

Containers are used to collect ash from the combustor (bottom ash), from the heat exchanger (top ash), and from the pollution control system (fly ash). Wood combustion ash is primarily composed of mineral oxides, and is valuable as a plant fertilizer or for other uses.

7.2.7 Exhaust Stack

The exhaust stack directs combustion gases up and out of the facility, where they are discharged to the atmosphere. An "induced draft fan" is usually installed at the entrance to the exhaust stack. Its purpose is to maintain a steady pressure drop across the entire system, ensuring steady operating performance of the various components.

7.2.8 Additional Components

Wood combustion systems typically do not operate well at extremely low loads—combustion temperatures drop and combustion efficiency drops, resulting in emission of undesirable, uncombusted carbon and hydrocarbons. If the facility is expected to have periods of low thermal load, it is important to provide for this through the addition of a small "pony boiler" that runs during times of low load (i.e., during spring and fall for a building heating system). This can consist of either a small wood boiler, or a fossil-fuel based system.

Thermal storage, in the form of an insulated water tank, is a very cost effective means for improving the overall performance of a hot water building heating system, by providing extra capacity during times of high load, reducing the need for low load operation, and reducing the tendency for thermal shock to the boiler when thermal load suddenly changes. Steam-based systems are not as amenable to thermal storage—instead, variable load steam turbines are sometimes installed to temper sudden changes in load to the boiler. Industrial heating systems typically have very steady demands for heat, and therefore do not benefit from the addition of thermal storage or load compensation.

Lastly, site layout and integration with the overall facility must be taken into consideration, so that fuel delivery and other operations seamlessly integrate into the larger functioning of the facility that is being served by the wood energy system.

7.3 System Performance

Wood combustion systems can impact several aspects of facility operation, including logistics, comfort, health and safety, and economics. However, when system performance is discussed, it is usually in terms of either thermal efficiency (for heat only systems), or electrical efficiency and overall energy efficiency (for CHP systems). Emissions performance is also a common point of comparison in terms of system performance.

7.3.1 Thermal Efficiency

Thermal efficiency of a combustion system is equal to the amount of heat energy utilized, divided by the total heat available in the fuel. Note that fuel heat value can be expressed as a "higher heating value" that includes the energy of condensation for water vaporized or created during combustion, or as a "lower heating value" that does not include the energy of condensation. It is best to use the higher heating value when calculating thermal efficiency, but one should always indicate which heat value is used when reporting efficiency values.

$$\eta_{\text{th}} = 100 \times q_{\text{thermal}} / q_{\text{fuel}}$$

where

η_{th} = thermal efficiency of system (%)

q_{thermal} = output of useful heat (W)

q_{fuel} = heat content of fuel (W, =GJ/ton × ton/h × 3,600/1.0e9).

Fuel heat content varies more as a function of moisture content than wood species, and tends to be approximately 14 GJ/ton for green (40 % m.c.) wood chips (Ciolkosz 2010). As heat is removed from the hot combustion gases, usually via a heat exchanger, the gases cool. If they cool to the point at which the flue gas temperature drops below its dew point, condensation occurs, forming droplets of water, usually on the walls of the flue. These droplets tend to be acidic (due to chemical

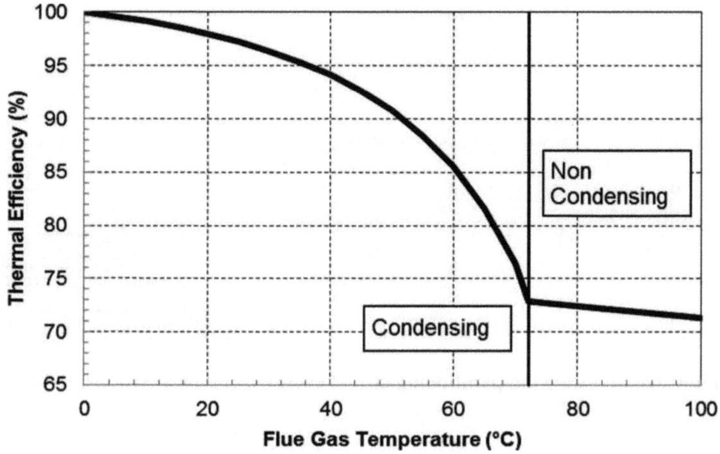

Fig. 7.5 Boiler thermal efficiency vs. flue gas temperature (after Neuenschwander et al. 1998)

reactions in the flue gas) and can be corrosive to components within the flue gas exhaust system.

As a result of the problems associated with condensation in the flue gas, most boiler heat exchangers are designed to only remove a portion of the available heat, so that the flue gases remain warmer than their dew point. Systems are usually designed to keep flue temperatures above about 150 °C to prevent condensation in the heat exchanger, as well as in the flue. As a result, system efficiency (controlled by the amount of heat removed by the heat exchanger) is limited to a maximum of about 70 % (Neuenschwander et al. 1998). Manufacturers wishing to design a higher efficiency system can install a second "condenser heat exchanger" that removes additional heat from the flue gas, and collects the condensate. Figure 7.4 shows the modeled thermal efficiency of a boiler when different amounts of heat are removed from the flue gas (assuming ambient air temperature of 0 °C). Note that, if the flue gas temperature drops below the dew point of 71 °C, efficiency rises dramatically, but condensation begins to occur in the flue gas. Notice that non-condensing systems typically maintain their flue gas temperatures much higher than the dew point, since condensation is related not to the temperature of the flue gas, but the temperature of the coldest surface in contact with the flue gas. Therefore, flue gas temperatures are usually kept significantly higher, to prevent condensation from occurring on any colder surfaces in the flue (Fig. 7.5).

One beneficial side effect of these condensing systems is that they tend to remove particulates from the stack as well, thus reducing emissions (Obernberger 1998). These systems must utilize stack components that are resistant to corrosion, such as high temperature polymers or corrosion resistant coatings. Thus far, condensing systems are not common in the commercial wood combustor industry in the Northeast USA.

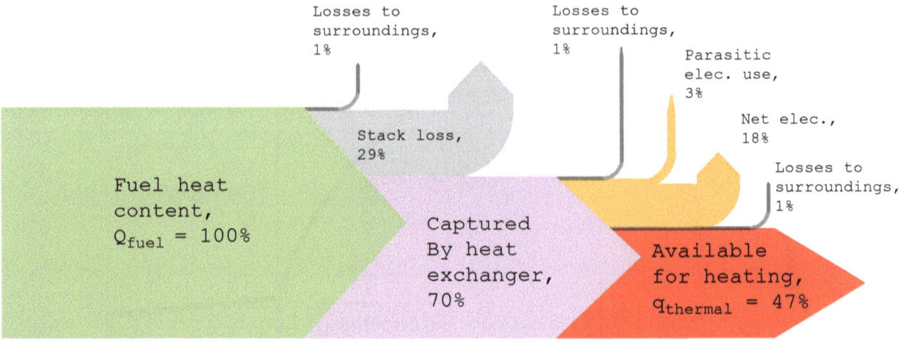

Fig. 7.6 Flow diagram showing partitioning of fuel energy in a typical Wood CHP System

7.3.2 Electrical Generation Efficiency

The efficiency of the electrical generator in a CHP system is equal to the rate of electricity produced, divided by the amount of energy delivered to the generator:

$$\eta_{\text{elec}} = 100 \times q_{\text{elec}} / q_{\text{steam}}$$

where

η_{elec} = electricity generation efficiency (%)

q_{elec} = rate of electricity generation (W)

q_{steam} = heat content of steam (W)

The electrical conversion efficiency is constrained by the thermodynamic limits of the generation cycle, but is usually in the range of 30–35 % for these relatively small systems.

In the case of a CHP system, the heat that is not used by the electrical generator can then be used to provide heat to the facility. As a result, a much higher % of the fuel's heat can be utilized. The overall efficiency of the system is equal to the amount of useful electricity generated plus the amount of useful heat generated, all divided by the amount of heat in the fuel.

$$\eta_{\text{total}} = 100 \times \left(q_{\text{elec}} + q_{\text{thermal}}\right) / q_{\text{fuel}}$$

where

η_{total} = overall energy use efficiency for the system (%)

q_{elec}, q_{thermal}, q_{fuel} defined above.

Efficiency losses from the system are primarily due to heat in the stack gases, but also include conductive/convective losses from the boiler and piping to the surroundings, and parasitic energy use by the system (Fig. 7.6).

If a high percentage of the heat leftover from electricity generation is utilized, the overall efficiency of the CHP system can be quite high—60–70 % is not impossible (the system depicted in Fig. 7.1 has η_{total} = 65 %).

7.3.3 Parasitic Loads

One of the more overlooked yet significant issues related to wood combustion and CHP system performance is that of parasitic loads, and their impact on overall system efficiency. Parasitic loads are those energy requirements (usually electrical) that are needed for operation of the combustion equipment. Blowers, belts, augers, pumps, and other equipment all require energy to operate, and this energy use reduces the overall efficiency of the system. It is not unheard of for a wood heat CHP system to have parasitic loads equal to over 15 % of the facility's electrical generating capacity (depending on system design and operation). Careful design of equipment to minimize parasitic loads can have a significant impact on operating costs and overall profitability of a facility. In the case of the example Wood CHP system shown in Fig. 7.1, $\eta_{elec} = 30\ \%$ if we ignore parasitic losses but drops to 26 % if we count them. In general, parasitic losses should be included when calculating system efficiency.

While the relatively high cost of electricity in some areas may make power production from wood appear to be an attractive business proposition, the reality is that, in order to be economically successful, there almost always must be a need for both the heat energy and the electrical energy. This is partly due to the fact that, even though heat energy is usually much less valuable than electrical energy, the amount of heat energy available from a CHP system is much greater than the amount of electrical energy that is produced. In the case of the sample system shown in Fig. 7.1, only 18 % of the fuel energy is available as electrical power, but up to 65 % of the fuel's energy can be utilized if both heat and power are put to use. Parasitic losses should be expressed both in terms of kW demand and as a percentage of the total fuel heat content.

7.3.4 Emissions Performance

Wood combustion emissions can include combustion products (CO_2, water), uncombusted wood particles (carbon and hydrocarbons), secondary gaseous products such as NOx and SOx, and ash particulate matter. While the public perception of wood combustion emissions sometimes includes smoky soot, commercial wood systems tend to be visibly clean with no odor or observable emissions from the stack. The actual levels of emissions can vary depending on combustion equipment design and operation, fuel type, and the amount of pollution control equipment used. Typically, emissions are reported in terms of mass of emitted material per GJ of fuel use (Table 7.1). See Chap. 10 of this text for a thorough discussion of the question of health impacts from wood combustion emissions.

Electrostatic precipitators (ESPs) can be used to reduce particulate matter emissions in locations where especially strict control is necessary—ESP performance tends to vary depending on equipment operation.

Table 7.1 Typical emissions rates for commercial wood combustion systems (grams per GJ of fuel energy)

System	PM (g/GJ)	PM2.5 (g/GJ)	NO*x* (g/GJ)	CO_2 (% by volume)	CO (g/GJ)
Direct combustor, no mitigation equipment	86–100	40–80	60–100	8–10	40–60
Direct combustor, multiclone	40–60	34–56			
Direct combustor, multiclone plus bag house	4–20	4–20			

Source: Babcock (2013); Hinckley (2008)

7.4 Emerging Technologies

While wood combustion systems are not a new technology per se, several new technological developments are promising to provide enhanced system performance in some scenarios.

7.4.1 Gasification Combustion

In spite of its inherent benefits for emissions and potential for gas turbine power generation, gasification combustion for wood has yet to emerge as a large component of the industry. Several technical and economic challenges continue to face wood gasification, including (among other things) the difficulty of excluding tars from the gas stream, and the challenge of working effectively with nonuniform feedstock. As the technology continues to develop, gasification systems may become an attractive option for wood combustion and CHP systems.

7.4.2 ORC

An Organic Rankine Cycle (ORC) system is a device that generates electricity using the same approach as a steam turbine, but using an organic fluid instead of steam. The organic fluid is completely recycled within the system, and can operate at lower temperatures than a steam-based system. This allows ORC devices to run off hot water or other "low grade" heat (albeit at a lower generation efficiency, due to the lower operating temperatures). One advantage of ORC systems is that, by avoiding the use of steam, requirements for onsite supervision of the facility are often reduced, relative to steam-based generation equipment. ORC systems have yet to gain a significant share in the wood CHP market, in part due to high installation and operating costs. As the technology develops, ORC systems should become increasingly economical for use.

7.5 Key Issues and Opportunities for Sustainable Growth of Combustion and CHP Systems

While current wood combustion and CHP systems are generally well designed and effective, there are several things that can be done to make commercial wood heat and CHP systems more attractive for expanded long term use in the region.

7.5.1 Installation Cost

When compared on a cost per gigajoule basis, wood energy is very competitive with both fossil fuels and other renewable energy. However, the cost of installation for wood systems is among the highest (see Table 7.2). When retrofitting existing facilities with wood heat, simple paybacks on the order of 5–7 years are common, and can be lower in some situations. However, if wood heat is to become a larger part of the market, reducing this initial installation cost is an important goal.

It is important to note that the costs shown above are only a portion of the total project. Total project cost for wood combustion and CHP systems is often dominated not by the cost of the combustor/boiler, but by the cost of the custom-designed support structures and interconnection equipment—especially when implemented in existing facilities. Part of this is due to the high cost of retrofit construction in the region, and part is due to "project creep"—the seemingly inevitable tendency for expensive extras and overdesign to be added in during the project development process. Many wood energy combustion projects in the Northeast USA have been supported by government grants. This has helped encourage the installation of equipment, but has also had an apparent effect of encouraging project creep, somewhat negating the benefit of the incentives.

A creative approach to system design and installation is needed to address this issue. For example, one approach could be to develop packaged, standard sized, skid mounted units that can be quickly and simply installed on site, and require a

Table 7.2 Typical fuel and installation costs for different systems

System	Fuel cost	Normalized fuel cost, \$/GJ	Installation cost (combustor/boiler/generator only)
Commercial wood heat	\$35–45 per ton	3.50–4.50	\$240–255 per kW
Natural gas	\$1.07 per therm	12.80	\$58–67 per kW
Fuel oil	\$4.08 per gallon	37.10	\$67–73 per kW
Electric boiler	\$0.13 per kWh	36.10	\$34–36 per kW
Commercial wood CHP	\$40–50 per green ton	3.50–4.50	\$700,000–750,000 per MWe capacity

Note: Data from DOE (2013); Babcock (2013); Mossman (2011). To convert GJ to Million BTU, multiply by 0.948; to convert from \$/MW to \$/Million BTUH, multiply by 293. Normalized fuel cost is adjusted for boiler efficiency minimum of interconnection to the building's systems. While some development has occurred in this area, it remains a significant opportunity for commercial scale wood energy systems in the region.

7.5.2 Automation

Initially, commercial wood combustion systems were used almost exclusively by the wood products industry—and with good reason. The availability of waste wood material, need for heat to run kilns, and availability of occasional labor in an industrial setting made wood energy systems a natural fit. However, other industrial and commercial facilities often do not have the same opportunities and constraints, and are more interested in an automated heat or CHP system that requires minimal effort to operate. Therefore, it is important for systems of the future to be better equipped with automated fuel handling, system restart, and ash handling systems, making them attractive to a larger segment of the population. Currently, labor requirements for a commercial boiler system tend to range from 1 to 10 h/week, depending on the size of the system and the degree of automation in place.

7.5.3 Emissions

Increased concern about air quality (especially aerial particulates) makes it essential that emissions control systems continue to develop. Current technology (i.e., cyclones, bag houses) is functional, but adds to both first cost and operating cost of systems. New, more cost effective approaches to emissions control would be very beneficial for the effectiveness and desirability of these systems. An emerging area of concern in the region is effective control of fine particulates (<2.5 μm in size).

7.5.4 Fuel Sensitivity

Fuel sensitivity is a significant technical challenge for wood heat systems, in terms of fuel size as well as the presence of dirt and other foreign material. Variable sized fuel can clog the handling system, while dirt and grime contribute to formation of klinkers (hardened chunks of ash) in the combustor as well as accelerated component wear. Both of these problems can reduce performance of the system and increase maintenance costs. These problems are a major obstacle to utilizing low cost wood material such as roadside tree trimmings, which are plentiful in some parts of the region. Innovative engineering design is needed to address these problems by creating equipment that is able to effectively utilize a wide range of wood fuels.

7.6 Concluding Comments

The use of commercial wood combustion systems is well established in the Northeast, and the systems have proven themselves to be economical and reliable. However, growth of the industry has been slow in spite of the economic and other benefits that derive from using wood fuel. Continued growth is likely to occur in the coming years, but the rate of the industry's growth will depend on several factors. Typically, commercial wood combustion systems have been used only for retrofits rather than new construction—thus there is an opportunity to challenge the architectural engineering design community to embrace wood energy as a desirable option. In addition, the degree to which commercial scale wood heat and CHP can expand will also depend on the continued development of combustor and fuel handling technology to make the devices more cost effective and convenient for the end user.

References

Babcock J (2013) Unpublished data. Advanced Recycling Equipment Inc., St Marys, PA

Ciolkosz D (2010) Characteristics of biomass as a heating fuel. Penn State Extension Renewable and Alternative Energy Fact Sheet. # UB043. The Pennsylvania State University, University Park, PA

EIA (2012) Annual energy review 2011. United State Energy Information Administration. Report #DOE/EIA-0384(2011) September 2012

EIA (2013) United States Energy Information Administration online fuel price data for the Northeast States. http://www.eia.gov. Accessed Jan 2013

Hinckley J (2008) Stack test report review for Bitter Root RC&D Area Inc. Resource Systems Group Inc., White River Junction, VT

Jenkins B, Baxter L, Koppejan J (2011) Ch 2 biomass combustion. In: Robert CB (ed) Thermochemical processing of biomass: conversion into fuels, chemicals, and power. Wiley, New York

Johnson E (2009) Goodbye to carbon neutral: getting biomass footprints right. Environ Impact Assess Rev 29(2009):165–168

Koppejan J, Loo S (eds) (2008) Handbook of biomass combustion and cofiring. Earthscan Publishing, London

Mossman M (ed) (2011) R.S. means mechanical cost data, 34th annual edn. Reed Construction Data, Norwell, MA

Neuenschwander P, Good J, Nussbaumer T (1998) Combustion efficiency in biomass furnaces with flue gas condensation. Biomass for energy and industry. Tenth European conference and technology exhibition, Wurzburg, Germany, 8–11 June 1998

Nussbaumer T (2003) Combustion and co-combustion of biomass: fundamentals, technologies, and primary measures for emission reduction. Energy Fuels 2003(17):1510–1521

Obernberger I (1998) Decentralized biomass combustion: state of the art and future development. Biomass Bioenergy 14(1):33–56

Whitmann T, Lehmann C (2011) Systematic under- and overestimation of GHG reductions in renewable biomass systems. Clim Change 104:415–422

Chapter 8
The History of the Wood Pellet Industry on the East Coast

Bruce Lisle

I can't tell the story of pellets on the East Coast without a brief digression on the residential pellet market development that began out in the West in the 1980s. The State of Oregon was the major instigator of the industry as they were the first to begin regulating wood stove smoke emissions, or what I term "the social costs of alternative energy." After the "oil crisis" that started in 1973, it seemed that anyone who could weld became a wood stove manufacturer. The use of wood for home heating had been on a constant decline since the end of World War Two, but the 1973 shortage reversed the decline. By the time we got into the 1980s, wood smoke was contributing to "brown cloud" in various air sheds, particularly in the West. Oregon, followed by the state of Washington, began to require wood stoves sold in their states meet certain emission standards. It didn't take too long for the EPA to pick up on this, and by 1988 all wood stoves sold in the USA were required to meet certain standards. The second step of this regulation, effective in 1990 required non-catalytic stoves to emit no more than 7.5 g of smoke particulate per hour. Testing costs dramatically reduced the number of wood stove manufacturers. In 1978 in Sandpoint, Idaho some entrepreneurs build the first wood pellet plant in the USA to provide fuel for commercial applications. A fellow by the name of Collins developed the "Collins Hopper" which was a pellet burner-hopper configuration that could be installed on a wood stove. A couple other inventors entered the fray, Joe Traeger from Mount Angel, Oregon and Jerry Whitfield of Everett, Washington. Joe developed the Traeger stove, furnace and pellet grill and eventually sold the stoves to Earth Stove, and the furnaces to Pinnacle in BC. Jerry's company, Pyro Industries, developed the Whitfield line of stoves and eventually sold to Lennox. Lennox eventually purchased the Earth Stove business.

The EPA exempted pellet appliances from the 1990 New Source Performance Standards (NSPS) regulations due to their high air to fuel ratio. Pellet stoves also

B. Lisle (✉)
Energex University, 600 Airport Drive, Mifflintown, PA 17059, USA
e-mail: bruce@pellet.pro

M. Jacobson and D. Ciolkosz (eds.), *Wood-Based Energy in the Northern Forests*, DOI 10.1007/978-1-4614-9478-2_8, © Springer Science+Business Media New York 2013 only emitted between 0.5 and 2 g/h and became known as the clean way to burn wood. Up until the mid to late 1990s, many different companies introduced new pellet stoves and the variety of designs was wide. But there was one general problem with these first generation stoves: they were designed to burn Western fuel. Most of the fuel made in the West was softwood from the spruce, pine, fir categories. These Western wood species picked up very little mineral content from the soil, as such; the ash content was in the 0.25–0.35 % range. This was great for the Western markets, but the industry was about to jump off of the Rockies and land on the Appalachian, Green and White Mountains in the East—hardwood country.

The first pellet plants built on the East Coast were actually in Canada. BioShell, a unit of Shell Petroleum, built three 125,000 ton/year plants, two in Northern Ontario (Hearst and Iroquois Falls) and one in Lac-Mégantic, Québec. These plants were built in the late 1970s and early 1980s. Today only the Lac-Mégantic, Québec operation survives. The first one in the Northeast USA was in Vermont (Green Mountain Pellet) and it made a cardboard pellet, operated by Pat Irish. The next one was in Virginia, Modular Energy Corporation. However, this is not where the market started. In the late 1980s, I was developing a combination pellet plant and cogeneration facility in North Sydney (Cape Breton), Nova Scotia. My company was known as Mapleleaf Technologies.

While I was in Cape Breton, we first purchased fuel from Modular Energy (MEC) in Virginia and trucked it to Nova Scotia. They had delivery problems which forced us to source our pellets from Montana. But MEC was selling fuel in the mid-Atlantic region. One of their customers was Courtland Hardware in the Baltimore area. Courtland was an early believer in pellet technology and became the largest dealer of Whitfield stoves. Richard Thomas, one of the owners of Courtland shared a story about MEC. A building regulation in the area around MEC's plant, required new homes to use fireproof plywood as underlayment on roofs. Eventually a defect with this building material required its removal and replacement. So now this large waste stream became available, MEC picked it up and made it into fuel. Richard had a customer who complained that MEC pellets he sold him wouldn't light. Richard, a hands-on retailer, made a house call. He couldn't get them to light either. After a call to MEC, it was determined they used *fireproof wood* to make fuel. Some people question why we need fuel standards. This was the early days.

So we shipped the first railcar of pellets in North America from Eureka, Montana to Cape Breton to feed a demonstration project of Whitfield and Earth stoves along with Traeger furnaces. What a mess we found when we opened the first railcar door! As the Cape Breton plant looked like a sure thing, around 1990 we decided to seed the Northeast USA with fuel to get the market started. So box car loads of fuel from Eureka were shipped to Palmer, Massachusetts where it was transshipped onto trucks for distribution. We lost money on each ton, but the desired effect was to get the Eastern market started, and that was the market we were going to pursue from Cape Breton. Developing a major industrial plant in Cape Breton soon became an insurmountable problem, even though we had the money raised, provincial government participation was required for a minor part. It didn't happen, so I eventually purchased the Lac-Mégantic, Québec BioShell plant in 1993 which turned out to be a much better deal.

At that point, eastern pellets were going almost exclusively to institutional and industrial customers. For example, Green Mountain Pellet was selling their cardboard fuel to institutions and even reselling BioShell's bark fuel from Lac-Mégantic. Hardly anyone was in the Northeast residential market yet. At 8.00 p.m. on 5 March 1993 I signed the purchase and sale agreement in Toronto to buy the Lac-Mégantic plant from BioShell. At 9.00 a.m. the next morning, I was setting up a booth at the Hearth Products Association Expo in Nashville. This was Darryl Rose (our VP of Marketing)'s first day on the job. We were sure that we had a jump being the first one in the Northeast market. But as we were setting up our booth in the far corner of the convention center, we looked down the aisle, and saw Catamount Pellet from North Adams, Massachusetts and Dry Creek Products from Arcade, NY. I had one of those Butch-Cassidy-and-the-Sundance-Kid moments, "who are those guys." The three of us were the first to establish commercial mills for the Northeast residential bag market. Averill Cook from Catamount and Barb Shine from Dry Creek, both outstanding individuals, eventually became Pellet Fuels Institute (PFI) presidents. And so did I.

In 1994, the Pellet Fuels Institute emerged as the merger of the Fiber Fuels Institute from Minnesota and the Association of Pellet Fuel Industries from the state of Washington. PFI was initially based in Minnesota and Pat Irish from Green Mountain was the first president, followed by Bob Massingill from California Pellet Mills, an equipment manufacturer.

By the mid-1990s, a slew of new mills started coming online. Steve Walker, who started with a 150 hp pellet mill in Framingham, Massachusetts, developed the Jaffrey, NH mill. Wood Pellet Inc in Summerhill, PA started along with Allegheny Wood Pellet in Youngsville, PA. Pennwood Products of East Berlin started making fuel at their molding plant from equipment they bought from Bear Mountain in Oregon. A couple mulch guys started Juniata Wood Pellets in Mifflintown, PA. Hamer Pellet started their first plant in WV along with American Hardwood Pellet in Upshur County, WV. There was also a very small operator, Renewable Energies Inc, in eastern West Virginia run by Jay Otto. Many of us built our own pellet plants, but Jay literally built his own pellet mill, the machine. He was in the hills of WV just over the border from Virginia along the I81 corridor. Vermont makes a living off of the leaf peepers during the Fall, but WV is the hidden jewel during the Fall. The hardwoods changing colors are beautiful. He told me that that whole area of WV was softwood when the first settlers got there, but it was all harvested to build the coastal cities in the region. And it all came back in hardwood, natural regeneration! By the late 1990s, Lignetics from Idaho developed a plant in Gilford, WV after a brief stop in Missouri.

On the Canadian side, Nova Scotia got its first mill built on the mainland near Halifax by Shaw Resources. Colonial, another molding plant, started producing small quantities from their waste in New Brunswick. Quebec was a free for all. As we first started the residential market in Québec, we were pleasantly surprised how this market quickly latched on to pellet stoves. Réal Nadeau, the husband of one of our office staff, worked for a couple distributors, and sold hundreds of stoves. Pellet mills sprang up all over the province. Granules LG in St. Félicien, Cubex made fuel from their parent company's flooring plant in Papineauville, Québec.

Fig. 8.1 From left to right: Bruce Lisle, Averill Cook, Bob Massengill and Barb Shine, leaders of PFI as it moved in with HPA

There was one small operator, to remain nameless, who burnt down, rebuilt and burnt down again. In Palmerston, Ontario, Langs Dehy (a feed mill), started early on to make wood fuel, but eventually dropped out a few years ago. I may have left out a couple operations, but these were the early Eastern pioneers.

One of the most devastating events of the period was the Hearth, Patio and Barbecue Association (HPA) Expo in Las Vegas in 1995. All of us Eastern producers were still ramping up production. This market started basically from scratch, and we had invested millions building these plants on the premise: *Build It and They Will Come*. A few market factors were starting to come to light, the big one being the higher inherent ash content of Eastern fuels. Bigger ash pans and more fuel friendly appliances were the order of the day. What was on the show floor was far from what we needed to develop this market. I remember the ash pans in most stoves were no bigger than an ash tray in a midsized Chevrolet. Fuel friendly? Forget it. But now what had been originally stand-alone pellet stove companies were developing gas units. Some of the leaders in the sectors jumped on the gas bandwagon. It seemed that if you were a dealer and you wanted to continue to sell a company's pellet stoves, you had to buy their gas fireplace as well, or the dealer across town would end up with both lines. I nicknamed this Expo, the Las Vegas Gas Show. Gas was the big presence, and pellet fuel producers were dismayed (Fig. 8.1).

When I got back from the show, I immediately designed these buttons, *Wood Pellets, No Go Boom*, and wore them at any event I subsequently attended (Fig. 8.2). Starting with the PFI conference right after the Las Vegas Gas show, producers banded together and concentrated on collective marketing the "Pellet Option" with specialty retailers. By the next HPA Expo in Charlotte, the pellet industry was on a

Fig. 8.2 Pellet Marketing Logo From the 1990s

mission. We collectively raised funds to make Charlotte a big event. We had the "Barb Shine Show." In a very large meeting room, we enticed dealers to attend by giving away truckloads of fuel, stoves, and even cash. We had several hundred people in the room. Barb got up on stage and preached the pellet option. We needed this exposure. Walking into the show the first day I was behind two dealers planning their day on what to see. One said he wanted to look at pellet stoves, the other said, "don't bother, that line is dead." They got a lecture from me along with an invitation to the Barb Shine show. From that point forward, the PFI has worked collectively to market the pellet option, always making the Expo a big event. We needed to expand the marketplace.

Another episode began in Charlotte. The PFI was based in Minneapolis, our Executive Director was working on a federal grant on industrial waste which was paying the bills of the association. But the association was now dominated by residential market fuel producers who wanted PFI to enhance residential markets. A couple of PFI Directors, including me, had discussions with Carter Keithly, the then President of HPA, about affiliating with HPA. Formal discussions followed these informal discussions, and late in 1996, PFI moved to Arlington, Virginia. We used HPA staffers to handle our day-to-day affairs, Don Kaiser was our next Executive Director, and we kept our own Board and financial structure. It worked well, and we had a Washington, DC presence, reduced overhead, and our members no longer had to pay dues to PFI/HPA/HPA regional affiliates. We now had more resources to collectively promote the pellet sector. One of Carter's motivations was that the HPA was getting dominated by gas, and Carter remembers HPA started as the WHA, the Wood Heating Alliance. He wanted renewables to be part of the equation (Fig. 8.3).

Collective marketing dominated PFI efforts until about 2004. We had seen the Y2K problem come and go, and we had a couple spikes in fossil fuel prices which always spiked the pellet sector. 2000 was the only year the industry ever saw fuel volumes decrease. About this time, fuel volumes in the East started to exceed

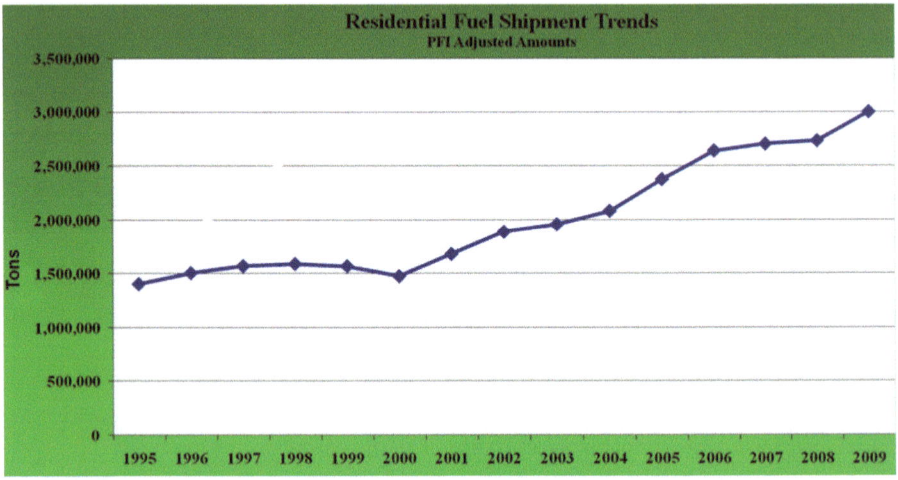

Fig. 8.3 Shipments, Adjusted PFI Data

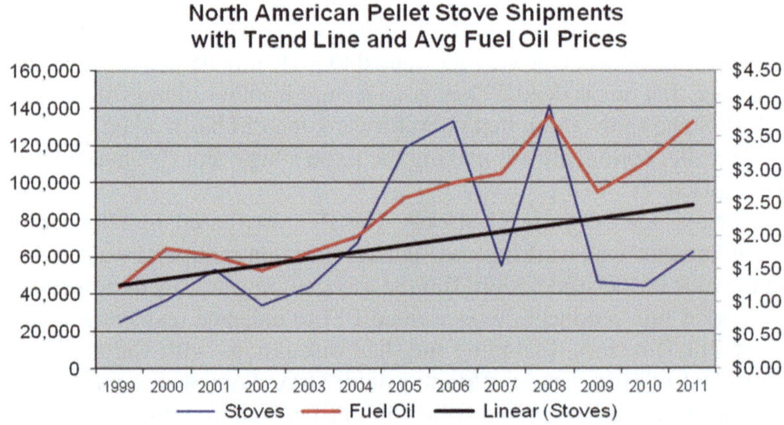

Fig. 8.4 Oil Prices vs. Pellet Stove Sales

volumes in the West, and the East continues to dominate the industry. Fuel oil and propain (that's what we call it) prices are the market drivers. The Northeast is so dominated by fuel oil, and it's a natural market for pellet fuel (Fig. 8.4 and 8.5).

One of the things I like about the pellet industry is that it developed based on free market economics. The industry has based its investment decisions on supply and demand, not slurping at the government trough. Over the last 20 years we have experienced shortage years, where fuel producers can't keep up with demand Fig. 8.5. It's these years which motivate entrepreneurs to build new capacity. If you look at the chart below, in a regulated environment such as electricity generation, the demand line would always be below the supply line. Utilities always have reserve capacity. In the pellet industry, capacity is build based on entrepreneurs investing in

Fig. 8.5 Editorial Humor

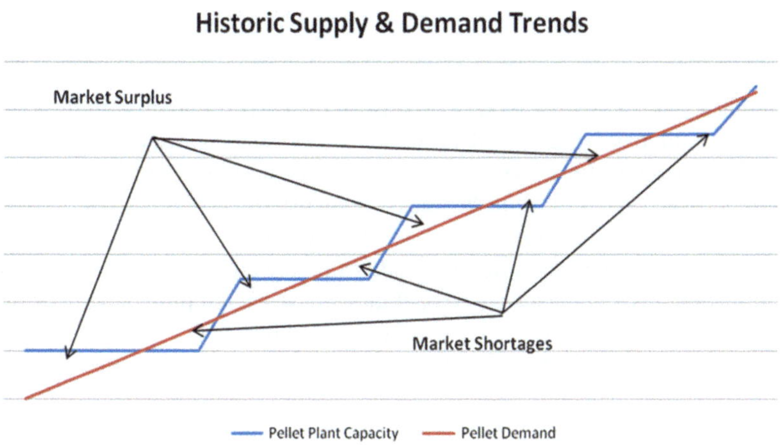

Fig. 8.6 Pellet Capacity & Supply-Demand

new plant. The motivation is usually a shortage year, so within 12–24 months, new capacity shows up and it takes market growth to absorb it (Fig. 8.6).

New capacity came on after Hurricaine Katrina in 2005, and most recently after the pellet shortage in 2008. Since the 1990s we had a few new plants in PA, New England built a couple in NY, Energex re-doubled its capacity again, Maine has three plants under way, Curran in NY. The mid-Atlantic region has seen new plants in West Virginia and Virginia—It's the capacity built after the 2008 shortage which is really a factor in the market today.

Fig. 8.7 Energex's Pellet Express Delivery Truck

In the last few years, the biggest expansion of pellet production in the USA has been in the Southeastern USA. Millions of tons of new capacity are being built in this region, almost all for export to Europe. Europe is subsidizing a fuel switch for its utilities, from coal and oil to biomass. It easier to transport and store biomass in the form of a pellet, so this has been the fuel of choice for the Europeans. The Southeast wood basket is rich in southern pine which is an agricultural crop in the south. High costs and excessive regulations have driven pulp and paper production offshore, so this excess fiber is making its way into pellet fuel.

The PFI has finished its fuel standards program and pellet mills are starting to get their fuel certified. The program has gone through many iterations, but the new NSPS regulations will now include pellet stoves. As such the US EPA is looking to have a consistent fuel to be used in certification process, which was the driving factor into adding third-party audits of fuel producers. It is also a sign of a maturing industry (remember MEC's fireproof plywood fuel). The consumers will now have a comfort that what is marked on the bag is in the bag (Fig. 8.7).

One of the most exciting developing markets for pellets is central heating. Consumers using fuel oil are currently paying the equivalent of about $470/ton pellets. Bulk delivery of fuel can cost less than half of this. There are a few manufacturers of furnaces and boilers. Many are imported from Europe, but the dominant US manufacturer is Harman Stoves in Pennsylvania. Once you add in what I call the middle market: commercial, light industrial, confined animal feeding operations, greenhouse and institutions, we anticipate that this market will soon dwarf the bag market.

Chapter 9
Biofuel Production

G. Peter Van Walsum and M. Clayton Wheeler

9.1 Introduction and Overview of Conversion Methods

Depleting fossil fuel resources and increasing CO_2 emissions have motivated the search for suitable biofuels to replace fossil derived fuels, particularly for the transportation sector. Because liquid hydrocarbon fuels such as gasoline, jet fuel, and diesel have high energy density and simple storage and combustion properties, they have provided an ideal solution to powering transportation. The problem of powering aircraft is particularly constrained since it combines the need for high energy content, small volume, low weight, tolerance to wide ambient temperature variations, and compatibility with simple propulsion systems. The search for renewable and low-carbon liquid fuels has resulted in the development of ethanol and biodiesel as renewable replacements for gasoline and diesel, respectively. While these fuels work in many applications and have lower carbon impacts than fossil fuels, to date they have both been produced primarily from food crops, either sugar or grain or vegetable oils, and as such compete with growing food demand. Responding to this need, there has been much research on developing biofuels that can be derived from cellulosic feedstocks, which can be produced more abundantly, less expensively and more sustainably than food crops. This chapter presents an overview of some biofuel technologies that could be appropriate for converting wood into liquid fuels.

G.P. Van Walsum (✉)
Chemical and Biological Engineering, University of Maine,
5737 Jenness Hall #311, Orono, ME 04469, USA
e-mail: PvanWalsum@umche.maine.edu

M.C. Wheeler
Chemical and Biological Engineering, University of Maine,
5737 Jenness Hall #307, Orono, ME 04469-5737, USA
e-mail: cwheeler@umche.maine.edu

M. Jacobson and D. Ciolkosz (eds.), *Wood-Based Energy in the Northern Forests*,
DOI 10.1007/978-1-4614-9478-2_9, © Springer Science+Business Media New York 2013

An important issue when considering biofuel production is the appropriate scale of the processing facility. In the oil industry, refineries operate at very large scale. The average refinery in the USA processes about 1.8 billion gallons of crude per year (US EIA 2013). Some refineries can reach scales an order of magnitude greater than this. Currently the largest corn ethanol mills in the USA operate at about 100 million gallons per year—an order of magnitude smaller than the average refinery. This scale is comparable to a pulp mill processing 3,000 dry tons/day of wood, and is thus near the practical limit for delivery of forest biomass to one location. Thus, identifying conversion technologies that can operate commercially at scales considerably smaller than an oil refinery is a key consideration for production of biofuels from wood.

9.1.1 Sugar Platform

One common approach to producing biofuels from wood is to use the carbohydrate content of the lignocellulose as a starting material for conversion to liquid fuels. The release of sugars from lignocellulose requires the hydrolysis of the cellulose and hemicellulose fractions of the wood, which for most species will yield glucose, xylose, arabinose, mannose, and galactose in varying proportions. These sugars can be converted biologically through fermentation, or converted using chemical processes. The fermentation route has been widely used to make corn ethanol and is compatible with the scale of regional biomass collection volumes. The challenge for deriving alcohols from lignocellulose lies primarily in reducing the cost of the hydrolysis step (Lynd et al. 1999). In these approaches, the lignin content of the wood is generally not converted to liquid fuels and is often utilized as a boiler fuel to provide heat and power to the process, and possibly for sale as co-products.

9.1.2 Pyrolysis

Heating up wood in the absence of oxygen will thermally decompose the material into different molecular arrangements, producing solids (char), liquids (bio-oil), and gasses. Done rapidly, pyrolysis will generate more oils; done more slowly it will generate more char and gasses. To provide liquid fuels, the oils are the desired product, and much effort has been invested in learning how to maximize oil yield and quality from fast pyrolysis. Most processes envision using catalysts and added hydrogen to upgrade the quality of the bio oil such that it is compatible with the fuel characteristics of transportation fuels. At UMaine, a method of generating upgraded pyrolysis oil in situ during pyrolysis is showing promise as a means of simplifying the incorporation of additional hydrogen into the conversion processes (Mukkamala et al. 2012).

9.1.3 *Hybrid Platforms*

Some processes have been developed that combine the sugar platform and pyrolytic conversion. Two such approaches presented here include the MixAlco process and thermal deoxygenation (TDO). The former digests biomass using a mixture of anaerobic organisms, and then pyrolyzes the resulting metabolic products to ketones and oils that can be upgraded to liquid fuels (Holtzapple et al. 1999). TDO is a similar process that first treats the lignocellulose with acid hydrolysis and dehydration prior to the pyrolysis step, and then can also add further hydrotreating to yield high quality drop in hydrocarbon fuels.

9.1.4 *Gasification*

Partial combustion of biomass to produce an energy-rich synthesis gas is another route commonly applied to making liquid biofuels. Gasification will reduce biomass to a mixture of CO, CO_2, hydrogen, methane, and water, along with undesired residual tars and chars. Through either fermentation or chemical catalytic methods, the synthesis gas can be converted to alcohols and/or hydrocarbons. These technologies have been widely investigated, and several demonstration and small commercial plants have been proposed. A common consensus on the application of gasification is that it requires very large scale to be cost competitive, and as such is not highly compatible with forest the use of derived biomass. Accordingly, gasification will not be further discussed in the chapter.

9.2 Hydrolysis and Fermentation to Alcohol Biofuels

Ethanol and butanol are established fermentation products, and both are suitable for use as fuels in spark-ignited engines. Ethanol produced from starch or sugar is considered a "first generation" biofuel. Ethanol produced from lignocellulose is considered a "second generation" biofuel, because this will create less competition with food resources. Butanol is sometimes referred to as an "advanced" biofuel, since it can be blended at any ratio with gasoline, although its energy density is lower than gasoline and so it is not a fully "drop-in" fuel.

9.2.1 *Feedstock Conversion*

Fermentation can convert the carbohydrate content of wood, the cellulose and hemicellulose, to desired fuel products. Fermentative organisms are not able to convert the lignin in wood, which requires an aerated environment for microbial decomposition. Because of this, fermentation pathways usually make use of lignin as a fuel to generate steam and power for the conversion processes. Excess energy from the lignin, if there is any, is expected to generate power or heat for export.

Conversion of sugars by fermentation is theoretically a high yield process, since the microbes chosen to catalyze the conversion can be developed to demonstrate high selectivity for the desired fermentation product. In an anaerobic fermentation, typically 10 % of the carbohydrate fed to the reactor is used as a carbon source for growth of cells, while the remaining 90 % of the carbohydrate is converted to fermentation products such as ethanol and carbon dioxide. The common fermentation products ethanol and butanol are metabolic waste products, but still retain approximately 95 % of the original energy in the carbohydrate feed. Thus, optimal energy yields of fermented alcohols on carbohydrate fed is about 85 %, which in practice can be achieved in ethanol fermentations. Butanol fermentations tend to have lower yields, depending on many process and microbiological variables.

It is worthwhile to note that, while the public often associates methanol with wood, it is not usually considered a good candidate for biofuel production because of its miscibility with water and high toxicity.

9.2.2 Pretreatment: Hydrolysis

The carbohydrates in wood are not readily accessible to fermentation as are the carbohydrates in sugar or starch crops. The large cellulose and hemicellulose polymers must be broken down into mono or small oligo-sugars for effective conversion. This is done by hydrolyzing the β 1–4 linkages in cellulose—a challenging task that requires aggressive hydrolysis conditions (strong acid or specialized enzymes). Hemicellulose is typically easier to hydrolyze to small carbohydrates, but is more difficult to ferment in the subsequent bio-conversion because it typically contains high quantities of 5-carbon sugars, such as xylose and arabinose, which are not consumed by many microorganisms.

Hydrolysis is typically carried out either by acid or enzyme catalysis. Acid hydrolysis is effective and can convert high percentages of the carbohydrate polymers. Unfortunately it is not highly selective and undesirable reactions that degrade sugar products to furans or organic acids reduce sugar yields. In addition, cellulose requires high severity reaction conditions to break down into glucose monomers, while these same high severity conditions tend to be too much for the hemicellulose sugars and they can be substantially degraded under conditions optimized for cellulose hydrolysis. This dilemma has led to several different configurations for acid hydrolysis, which often involve multistep of flow-through reactor configurations that improve yield but add process complexity. Another disadvantage of acid hydrolysis is that the severe reaction conditions effective for production of sugars are expensive to operate. Low pH conditions engender high costs for purchase and final disposal of the acid and base chemicals and expensive metallurgy is required to withstand the highly corrosive conditions.

Enzymatic hydrolysis can address several of the shortcomings of acid-catalyzed hydrolysis. Enzymes are highly selective, and will generate sugars from carbohydrate polymers without producing unwanted side or byproducts. Enzymes also operate at mild process conditions, such as 55 °C and pH 4. Thus, use of enzymes can increase the yield of sugars recovered from wood carbohydrates, thereby lowering per-gallon costs for fuel. The principal drawback of enzymes is that they are expensive and that they operate very slowly. Enzymatic hydrolysis of cellulose can require days to reach high yields, as opposed to acid hydrolysis which would typically operate in the tens of minutes time span. There has been intense research activity over the last decade to improve the effectiveness of enzymes, and much progress has been made. However, even the most advanced enzymes can only achieve optimal performance if the biomass feedstock has been adequately pretreated prior to the hydrolysis. Without pretreatment the rate of enzyme hydrolysis is so slow and the yield of sugars so low that the process becomes completely untenable.

A broad variety of approaches to pretreatment have been tested for their ability to improve rates and yields of enzyme hydrolysis. Many proposed pretreatments consist of cooking the biomass in aqueous solution. The methods differ in the pH of the reaction and the temperature and duration of cooking. Low pH pretreatment tends to solubilize hemicellulose sugars and precipitate the lignin, making flow-through systems desirable to avoid redeposition of the lignin on the cellulose. Low pH pretreatments can be tuned to give high yields of sugars by minimizing sugar degradation while maximizing enhancement of enzyme activity. The major drawback of low pH systems is that they do generate some degradation products that are inhibitory to both the enzymes and the fermenting microbes. High pH systems tend to solubilize lignin but leave the hemicellulose largely in oligomeric form, requiring more active enzymes. Alkaline processes tend to produce slightly lower overall sugar yields, but typically the hydrolysate is less inhibitory to the biocatalysts than the low pH product. Neutral pH systems can work well, often making use of autohydrolysis resulting from release of endogenous acids from the biomass. Because the chemical environment is less severe under near-neutral conditions, the duration and temperature of the treatment are typically increased to compensate for the reduced chemical activity. This can again lead to generation of inhibitory compounds. In general, hardwood responds well to most aqueous pretreatments, while softwood is more recalcitrant and hard to work with. Typically, research groups focused on softwoods tend to favor SO_2-catalyzed pretreatment. Many other variants of pretreatments have been proposed: Use of solvents such as ethanol or acetone; supercritical fluids; wet oxidation; steam explosion; energetic systems such as exposure to ultrasonic waves or shock waves; use of ionic liquids; pulping processes; and combinations of different approaches. To date, most pretreatments that have been optimized for a feedstock are effective at enhancing enzymatic hydrolysis, but they are estimated to have high capital and operating costs. It has been said that "The only thing more expensive than pretreatment, is no pretreatment," meaning that enzymatic hydrolysis without pretreatment would be even more expensive. The very diversity of approaches to pretreatments suggests that there is no obvious choice for a technology that delivers high performance at low cost.

9.2.3 Primary Conversion: Fermentation

Fermentation can be used to convert sugars into a tremendous array of compounds. For the purpose of biofuels, anaerobic metabolic products such as ethanol, butanol or methane are suitable candidate products since they are produced in large volumes, contain high levels of energy and are primary metabolites, meaning they are the byproduct of regular cell growth, not a product of unusual culturing circumstances. Biological conversions are generally highly energy efficient, but like most thermodynamically efficient systems, they also progress at slow rates. Thus they require long residence times and large reaction vessels.

Like enzyme hydrolysis, fermentations are carried out under mild conditions, and a common approach to configuring these two process steps is to combine them together into one common vessel, the so-called simultaneous saccharification and fermentation (SSF). SSF reduces process complexity and also offers some kinetic advantages for the hydrolysis, since the sugars produced by the enzymes and consumed by the microbe inhibit the hydrolysis rate. When hydrolysis is carried out in the presence of the fermenting organisms, feedback inhibition by sugars is diminished. The drawback to SSF is that it usually requires a compromise of pH and temperature conditions between the different optimal conditions for cell growth and the enzyme activity.

Growth of pure culture fermentations in large volumes using wood-derived sugars presents several challenges that have yet to be overcome at a commercial scale. In addition to the high cost of hydrolysis and the resulting inhibitors, many naturally occurring organisms are not capable of fermenting 5-carbon sugars such as xylose, which are abundant in hardwoods and herbaceous plants. Efforts over the last 20 years have produced engineered organisms capable of fermenting the variety of wood-derived sugars to ethanol. Although these efforts have met with much success, in most cases the organisms are less robust than their parent stock and are more susceptible to inhibition by high levels of alcohols, organic acids, or wood-derived inhibitors. Thus, at this time, there still remains a need for an industrial quality organism capable of fermenting 5-carbon sugars to desired fuel products. A related issue is that of contamination. If cultures are slow growing because of inhibition or difficulty converting C5 sugars, this opens up the possibility of a contamination by more well-adapted, but non-fuel producing organisms. Even current industrial scale fermentations converting corn to ethanol can fall prey to contaminants, and antibiotics are often applied. It is likely that the more difficult fermentation of wood-derived sugars will also require some level of antibiotic protection against contamination.

In addition to the carbohydrate derived from the wood feedstock, fermentations also require additional nutrients to enable growth of cells. Wood alone is deficient in nitrogen and other required macro and micro nutrients. Acquisition of these necessary nutrients adds cost to the process, although some of the inorganic nutrients may be recyclable in the process. Finally, the waste streams from fermentation systems are very high in Biological Oxygen Demand (BOD), and adequate waste treatment facilities are necessary to accept the waste streams resulting from fermentations.

9.2.4 Additional Conversion Methods or Upgrading Requirements

For use in spark ignition engines, ethanol and butanol do not require additional upgrading beyond separation from the fermentation stream. Higher energy density fungible fuels can be derived from these alcohols through hydrotreating (see below).

9.2.5 Purification

Separation of ethanol from the fermentation broth is easily accomplished through distillation and dehydration. This is mature technology and provided that the fermentation broth has reasonably high ethanol content (at least 5 %), optimized distillation systems use a relatively small amount of energy relative to the energy content of the ethanol product. After distillation, dehydration using molecular sieves is also a well-established practice in the fuel ethanol industry.

Separation of butanol from fermentation broth is considerably more complex than ethanol. First, butanol is more toxic to the microbes than ethanol, and so the maximum concentration of butanol in the final fermentation broth is typically less than 2 %, requiring considerably more water removal. Second, butanol has a boiling point above that of water, making separation by distillation alone far too energy intensive. Instead, many clever process designs have been developed to reduce the energy demands of separation, but they increase process complexity to do so. Solvent extraction can remove butanol directly from the fermentation. Once butanol reaches a concentration of about 20 % in water, it will phase separate, simplifying further water removal.

9.2.6 Description of the Product

Ethanol and butanol are combustible in spark ignited internal combustion engines, and have energy densities (HHV) of 29.7 and 36.0 MJ/kg, respectively. A benefit of biofuel production by fermentation is that the final product of the fermentation—an alcohol—is directly useable as a fuel without any need for further upgrading. Ethanol can be directly blended into gasoline at up to 10 % for cars driving in North America, and flex-fuel vehicles are currently on the road that can accept anywhere up to 85 % ethanol. The technology for burning neat (100 %) ethanol is also fully developed, though not currently deployed in the USA. Ethanol does pose the problem of not being a fungible fuel, meaning that it is not suitable as a direct replacement for gasoline. This is not so much a function of how ethanol burns in an engine, but is primarily because ethanol is fully soluble in water, which makes it very difficult to utilize in the pipeline distribution network currently used for liquid fuels. Butanol is more compatible with the gasoline infrastructure as it can be blended with gasoline in any proportion for use in spark ignited engines.

The metabolic yield of ethanol on carbohydrate is 51 % by mass, with 1 mol of glucose producing 2 mol of ethanol. The theoretical yield is typically 90 % of the metabolic yield, as some of the carbon in the sugar is required for growing the cells. Yield of sugars from wood carbohydrate can approach 90 % of the theoretical yield though this may involve unacceptably high enzyme costs. The remaining 10 % of carbohydrate either remains in its unfermentable polymer form or is degraded in the pretreatment process. On a mass basis, hydrolysis of carbohydrate polymers to monomer sugars adds about 11 % to the mass balance. Thus, an optimistic conversion of carbohydrate to ethanol would be: 90 % (yield of hydrolysis) × 111 % (stoichiometric mass of hydrolysis) × 90 % (fermentation yield) × 51 % (metabolic yield) = 45.8 %. On an energy basis (HHV), ethanol holds 29.7 MJ/kg, versus 17.4 MJ/kg for the starting holocellulose. At a 45.8 % mass yield, this results in an energy recovery of 45.8 × 29.7/17.4 = 78.2 % energy recovery in the final fuel product.

9.2.7 Application to Forest Resources

Technology for converting lignocellulose to ethanol is currently being deployed at commercial scale. In 2013, the US DOE declared that the major effort to develop cellulosic ethanol technology had succeeded: that pilot projects and techno economic models for converting cellulosic feedstocks to ethanol were predicting a minimum ethanol selling price of an "*n*th" plant to be on the order of $2.15/gal (DOE 2013). Concurrently, established biomass processing companies such as Dupont (chemicals), POET and Abengoa (corn ethanol), and Old Town Fuel and Fiber (OTFF, Wood pulp) announced that they were going ahead with commercial scale development of cellulosic ethanol production facilities (DOE 2013). Corn stover is the feedstock of choice for three of these companies, while OTFF intends to use hardwood as their feedstock. If the plans of OTFF are realized, they will be an early demonstration of a commercial venture in the Northeast to convert wood to ethanol. Current energy markets and future pricing are highly speculative at this time, as the advent of hydraulic fracturing technology is revolutionizing the fossil fuel industry in North America. However, if commercial ventures succeed in producing biofuels (and profit) from wood, then investment in this sector may increase. It is expected that the preferred early adoption wood feedstock would be hardwoods rather than softwoods because the former hydrolyzes more easily. Scale of fermentation biofuel plants would likely need to be similar to that of large existing pulp mills (~2,000 t/day) to be commercially viable. Some smaller niche applications such as that of OTFF may also prove to be viable.

9.3 Fermentation to Advanced Biofuels

Ethanol is an easily produced fermentation fuel. Currently it is the most abundantly produced liquid biofuel, with leading production coming from the USA and Brazil. However, its non-fungibility limits its practical usability in the USA. In addition, ethanol is not a practical replacement for diesel or jet fuel. What is really needed is an array of advanced biofuels that are fully compatible with petroleum gasoline as well as diesel and jet fuel. These are often referred to as "drop in" fuels because of their ability to be interchangeably swapped in place of petroleum-based fuels (i.e., they are fully fungible). Advanced biofuel production has yet to be demonstrated at an industrial scale in the Northeast, but shows promise if scientific and engineering developments continue.

9.3.1 Feedstock Conversion

As with alcohol fuels, advanced biofuels can be derived from microbial consumption of fermentable sugars, thus the lignin fraction of wood would most likely be separated out and used for heat and power. Unlike alcohol fuels, hydrocarbon molecules are most commonly generated by microbes growing under aerobic conditions, and they accumulate as energy storage molecules, rather than metabolic waste products. Growth under aerobic conditions progresses more quickly than anaerobic fermentation, and complete conversion of sugars is normally achieved. "Drop in" hydrocarbon fuel molecules can be derived from lipids extracted from the cells of the microbes, and the same technologies used for producing fungible fuels from vegetable oils (i.e., transesterification and hydrotreating) are applicable to the microbial lipids. Some microbes have been genetically engineered to produce a final desired hydrocarbon molecule without the need to upgrade or hydrotreat. These technologies have not yet been demonstrated at larger scale.

9.3.2 Pretreatment: Hydrolysis

For production of advanced biofuels, hydrolysis of the feedstock is essentially the same as for alcohol fermentations, though the more robust growth of the aerobic organisms may give them higher tolerance for inhibitory conditions.

9.3.3 Primary Conversion: Fermentation

Lipids accumulated by the microbes serve as energy storage compounds. When oleaginous (lipid-accumulating) microbes are cultured in an environment that is energy (i.e., carbohydrate) rich, but limited in a nutrient required for cell replication, such as nitrogen, they will cease allocating carbon to cell reproduction. Instead, they will divert energy for storage in high energy density molecules, such as lipids. Typically the microbes accumulate these lipids within their cell walls, so as not to lose their energy reserve to other organisms. Harvesting the lipids involves harvesting the cells and then releasing from their cell mass the desired lipids. There have been reports of cells accumulating lipids at very high densities, up to as much as 70 % of their cell mass. These accumulations were reported for cells growing under highly ideal laboratory conditions. Lipid accumulations under less optimal conditions, more representative of an industrial situation, have not achieved yields this high and figures in the 10–25 % range are more common. Achieving optimal lipid yield from the carbohydrate feed involves balancing the production of cells with the accumulation of lipids in cells. A common approach is to grow the cells in two fermentation stages. The first stage accumulates cells by providing all the nutrients necessary for cell replication, the second stage is a "fattening up" stage, in which carbohydrates are provided to the cells but not the growth-inducing nutrients.

Much of the work done on oleaginous organisms has focused on specific microbes selected for their ability to accumulate high quantities of lipids. Some work has also been done culturing mixed microbial communities in a feast/famine cyclical nutrient environment. Exerting this environmental pressure on a mixed culture of microbes will select for microbes with effective energy storage capabilities. This holds promise for eliminating the costs associated with maintaining aseptic conditions necessary for growth of a pure culture. The drawback of this idea is that different microbes use different energy storage molecules, many of which, such as poly hydroxyl alkanoates, are oxygen-rich and less energy dense than lipids.

Work has also been done to genetically engineer microbes that excrete hydrocarbon, or similar molecules, into the fermentation broth. Because they are hydrophobic, these molecules phase separate from water and can be readily harvested. Scale up of these fermentations has not yet been demonstrated, in part because of contamination issues. It is likely that organisms that freely give up their hard won energy reserve molecules would be uncompetitive with a fast growing contaminating organism.

Energy storing microbes grow in an aerobic environment. This allows them to extract more energy from the carbohydrate feed, but also results in greater loss of carbon to CO_2. Because they have more energy available to them, aerobic cultures grow faster than anaerobic cultures, and thus require shorter residence times and smaller fermentation vessels. However, providing oxygen to the microbes requires application of energy intensive aeration and mixing action, and also results in cooling requirements for the fermenters. Aeration, mixing, and cooling all add cost to the operations.

9.3.4 Additional Conversion Methods or Upgrading Requirements

Microbial lipids are chemically similar to vegetable oils, and conversion of lipids to useable fuels is accomplished using the same processes. Biodiesel (long chain methyl esters) can be produced by transesterification of the lipids with methanol, generating glycerol as a byproduct. The methyl esters can also be upgraded to oxygen-free hydrocarbons by hydrotreating. This has been demonstrated with vegetable and algal oils, and fuels meeting jet fuel specifications have been produced and demonstrated by commercial airlines. For microbes that produce hydrocarbon fuel molecules directly, no upgrading should be necessary, although these fermentations have not yet been successfully demonstrated at industrial scale.

A second approach to converting oleaginous organisms is to avoid the issue of separating the oils from the cell bodies and simply using the entire cell mass as a feedstock for fuel production. In addition to the stored lipids contained with the cell, the microbial cell wall itself is constructed of lipid molecules and thus is also an energy rich material. Recent efforts have demonstrated production of a biocrude oil by pyrolysis or hydrothermolysis of the entire cell mass. The biocrudes produced from these processes have not yet been fully characterized, but at least a portion of the oil is hydrophobic and thus a promising intermediate for hydrocarbon production. It is expected that upgrading these biocrude oils to hydrocarbons would present challenges similar to those of regular pyrolysis oils.

Alcohols such as ethanol and butanol can also be used as feedstocks for chemical upgrading to hydrocarbon fuels via well understood catalytic processes. Such upgrading consumes hydrogen and typically involves heterogeneous catalysts and pressurized reaction conditions. The resulting fuels have higher energy density and compatibility with the petroleum-based fuel systems, but because they no longer contain oxygen, the mass and volume yields of these fuels are lower than their alcohol feedstock. Although these upgrades are chemically feasible, the added value of the fuel has not to date prompted investments in the additional cost for the upgrading.

9.3.5 Purification

For cell-associated lipids, harvesting the cells from the culture broth is the first step in purifying the fuel product. Separation of cell mass from the fermentation broth is relatively easy, especially if larger microbes such as yeast are cultured. The greater challenge of the separation is the extraction of the oils from the cell capsule. All methods currently employed for this separation are energy intensive, operationally complex, or both. Cells can be disrupted mechanically or subjected to solvent extraction, but neither has yet shown promise for economical scale up. Oils derived from microbial cells are suitable for upgrading using conventional approaches used for biodiesel production. After extraction of the oil, the remaining cell mass, rich in protein and nutrients, is typically assumed to be suitable for sale as a supplement to animal feed.

For systems in which the organisms excrete their lipid or hydrocarbon product, product separation is envisioned to take advantage of the phase separation and is not expected to be a formidable challenge.

For processes opting to convert the whole cell mass, the main separation issue is likely to be drying of the material prior to pyrolysis. Treatment by hydrothermolysis will likely involve dewatering to concentrate the cells to a level of optimal water availability. It is possible that the resulting oil may contain elevated levels of nitrogen or other contaminants originating from the cell mass. This may require additional chemical clean up prior to using as a fuel.

9.3.6 Description of the Product

It is anticipated that production of advanced biofuels from fermentation processes will employ catalytic hydrotreating and hydrogen addition methods that will render the fuel sufficiently compatible with the hydrocarbon infrastructure. Advanced biofuels could be usable as a neat replacement for petroleum, or converted to a crude supplement that could be fed into an existing refining operation. However, the ideal is that new hydrocarbon fuel would be usable as a straight "drop in" fuel.

In general, aerobic fermentations release half of their intake carbon as CO_2, with the remainder going to cell mass. Cell mass itself, on a dry basis, is about 50 % carbon. If the lipids are 25 % of the cell mass and 80 % carbon, then their mass and carbon yields are about 10 % and 8 % respectively, on initial glucose.

With the introduction of oxygen into the conversion process, carbon dioxide, rather than alcohols, becomes the major metabolic waste product, converting a higher portion of incoming carbon to a non-energetic product. In addition, aerobic fermentation also accumulates more cell mass than anaerobic growth.

9.3.7 Application to Forest Resources

Production of hydrocarbons by fermentation is only starting to reach commercial application. To date commercial ventures have focused on producing higher value oils for fine chemical markets, and using low cost and easily fermented feedstock such as sugar from tropical sugarcane. As these technologies mature they may be able to reach scale that is large enough to support hydrocarbon production for biofuel markets. Eventual application of these technologies to fuel production from wood would likely focus on hardwoods, as they are more easily pretreated and hydrolyzed to fermentable sugars than softwoods. Fermentation is relatively easy to operate at biomass friendly scale, as is done in the corn ethanol industry, but the capital cost of pretreatment and hydrolysis to generate fermentable sugars would likely require facilities on the scale of 2,000 dry tons per day of wood.

9.4 Pyrolysis Platforms

9.4.1 Feedstock Conversion

Pyrolysis can convert both the carbohydrate and lignin content of wood to liquid fuel products at relatively high energy yields of 60–70 %. The process is simple, and equipment can be small-scale because the reaction times for the conversion are as short as a few seconds. In conventional pyrolysis, the wood is heated in the absence of oxygen to between 400 and 500 °C. The byproduct char can provide enough energy to both dry the feedstock and accomplish the conversion. The composition of the feedstock can significantly affect the product yields and quality. For instance, pyrolysis oils produced from woody biomass have higher oxygen content and have lower heating values than those produced from agricultural seed-oil residues, which contain more hydrocarbon-like starting components (Strahan et al. 2011). Pyrolysis methods are usually classified as either slow or fast depending on the rate at which the feedstock is heated. The primary product of slow pyrolysis is a bio-char, while the primary product of fast pyrolysis is bio-oil. In fast pyrolysis, the wood is heated and decomposes in only a few seconds. Over the past 40 years, researchers have been interested in converting bio-oil into petroleum-like hydrocarbons (Furimsky 1983; Elliott 2007). Fast pyrolysis requires high heat transfer rates which are accomplished by grinding the wood to particle sizes between 200 μm and 6 mm, depending on the fast pyrolysis technology. The particles are also dried to about 10 % moisture in order to minimize the water content of the water-soluble fraction of the bio-oil.

9.4.2 Primary Conversions

Fast pyrolysis reactor configurations include: vacuum moving bed, rotating cone, entrained flow, circulating fluid bed, ablative, transported bed (commercialized by Ensyn), with the most common technology being the fluid bed technology (commercialized by Dynamotive) (Bridgwater 2000). Common to all technologies is the need to separate char from the pyrolysis vapors and rapidly quench polymerization reactions by condensing the vapors, which can be difficult because they are primarily aerosols. Also required by all pyrolysis methods is a means of improving the quality of the oil by reducing oxygen and water content and increasing energy density and shelf life. These needs have been addressed through various approaches including catalytic upgrading and hydrotreatment of some kind.

Table 9.1 Comparison of lignin pyrolysis by conventional, lime-augmented, and formate-assisted pyrolysis (Mukkamala et al. 2012)

Feed	Lignin	Lignin/$Ca(OH)_2$	Formate/lignin,0.5 g/g	Formate/lignin, 1 g/g
C yield in oil (wt%)	21.3	20.4	23.4	28.6
O:C	0.19	0.21	0.14	0.067
H:C	0.96	0.97	1.23	1.40
HHV (MJ/kg)	30.7	32.1	37.2	41.7

9.4.3 *Pyrolysis Oil Upgrading*

Approaches for reducing the oxygen content of bio-oil include hydrodeoxygenation (HDO), catalytic pyrolysis, and their combinations. Hydrodeoxygenation uses high-pressure hydrogen and catalysts to convert oxygen to water, add hydrogen to the molecules, and crack the larger molecules to gasoline-range compounds.

Hydrodeoxygenation can either be done prior to condensing the pyrolysis vapors or after they have been condensed. The latter is more practical for distributed processing because the wood can be liquefied in the forest for transport to a centralized upgrading facility where the economics hydrogen production and catalyst regeneration benefit from a larger economy of scale. Examples of metals used for HDO catalysts include nickel, ruthenium, platinum, palladium, and cobalt/molybdenum catalysts commonly used for hydrodesulfurization of petroleum (Elliott 2007; Zhang et al. 2011). Common HDO concerns are the cost of purchasing or producing hydrogen, catalyst activity and stability, and product yield losses from coke formation on the catalysts and cracking of hydrocarbons to light gasses. In fact, condensed phase upgrading of a fast pyrolysis oil to gasoline can decrease the overall energy yield from 60 % for the bio-oil to 20 % or even lower for the gasoline product.

An alternative to upgrading pyrolysis oil that is currently being developed at the University of Maine is to blend the lignocellulosic feedstock with a divalent formate alkali salt, such as calcium formate. This addition of calcium formate to the reactor feed contributes hydrogen to the pyrolysis environment and results in production of bio oils that have substantially reduced oxygen content and increased energy density and stability. Table 9.1 presents experimental data comparing formate-assisted pyrolysis of lignin to conventional pyrolysis. It can be seen that the formate assistance improves yield, oxygen and hydrogen content (lower and higher, respectively) and results in higher heating value (HHV)—an impressive result given the simplicity of the method.

9.4.4 *Purification*

After successful hydrotreatment and stabilization, pyrolysis-derived oils should be amenable to distillation for fractionation and purification. Properly distilled, stabilized pyrolysis oils may be suitable as drop in or blend stock fuel agents.

9.4.5 Description of the Product

The weight fraction of oxygen in untreated pyrolysis oil is similar to that of the woody feedstock (40–50 %). This oxygen is responsible for the instability of the oil as well as the low heating value of the oil, which is about half that of petroleum crude oil. Untreated pyrolysis oils have limited use as boiler fuel because they have large water content (30 %), are corrosive, and they polymerize over short periods of days or months and become very viscous, or even solid at room temperature. The quality of FAsP oils is considerably improved over untreated pyrolysis oils, and can be produced without the need for additional catalytic upgrading or hydrotreating (Mukkamala et al. 2012). Highly hydrotreated oils would likely be designed for use either as drop in petroleum fuel replacements or blend stocks useful to crude oil refining operations.

9.4.6 Application to Forest Resources

Pyrolysis appears to be a suitable technology for scale down. Simple (non-upgrading) pyrolysis reactors have been developed that process on the order of 10 or 50 t/day of wood, and are small enough to mount on mobile trailers. Energy for the operation of these reactors is largely derived from the combustible gasses given off by the reactor, making good use of this low value byproduct. Technologies such as these may allow for primary processing to be distributed in the field and then feeding intermediate energy-densified products to central upgrading facilities.

Simple pyrolysis systems do not handle lignin very well, as it is largely converted to char products and high lignin content can lead to feed handling problems. Accordingly, lower lignin wood resources would likely be a preferred feedstock for pyrolysis operations, particularly if depending on relatively simple mobile reactor designs. As shown above, formate-assisted pyrolysis appears to render lignin more amenable to pyrolytic conversion to oils with desirable characteristics.

9.5 Carboxylate Platform to Alcohols and Hydrocarbons (MixAlco Process)

Fermentation technologies considered above make use of selected microorganisms converting the feedstock sugars to a single desired product. A liability of these fermentations is that they are highly susceptible to contamination. This can be particularly true for genetically altered organisms or organisms selected for traits that result in high yields of product at the expense of reduced cell replication. An alternative approach to fermentation conversion is to use mixed cultures of organisms that are cultivated under conditions that result on the accumulation of organic acids, a process known as acidogenic digestion. The process is similar to that of producing biogas through anaerobic digestion, but the growth of methanogenic organisms is inhibited, which results in an accumulation of organic acids. These acids then act as a feedstock for production of organic chemicals or fuels by means of chemical conversions.

9.5.1 *Feedstock Conversion*

As with other fermentation methods, acidogenic digestion does not convert the lignin in wood. However, since the fermentation includes a mixture of organisms, the microbial ecosystem will decompose all other materials that are digestible under anaerobic conditions, including hexose and pentose carbohydrates, proteins, and sugar acids. Higher energy fermentation products, such as alcohols or hydrogen, become intermediate products of the fermentation and are also consumed. The thermodynamics of fermentation dictate that if methane is not a possible final product, the metabolic conversions will terminate in the production of organic acids. Because the fermentation is non-sterile, rich biomass wastes such as manure or sewage sludge can be used to provide low cost nutrients. The resulting feed mixture of lignocellulosic materials and organic wastes is typically variable in composition and difficult to analyze, thus conversion is usually reported in terms of volatile solids consumption. Although strongly dependent on composition and fermentation conditions, a useful rule of thumb is that mass yield of organic acids can be expected to be on the order of 60 % of volatile solids fed to the digester.

9.5.2 *Hydrolysis*

In acidogenic digestions, hydrolysis of the feedstock is carried out by the mixture of organisms themselves, without the addition of supplemental enzymes. Thus, these systems can operate under conditions of Consolidated BioProcessing (CBP), in which the hydrolysis and fermentation of all carbohydrate materials are carried out in one vessel. As with other biologically mediated hydrolysis methods, pretreatment is very beneficial for acceleration of the hydrolysis process. Because the fermentations are acidic and require buffering, alkali pretreatments, such as lime or ammonia, are preferred over acidic or neutral pH pretreatments.

9.5.3 *Primary Conversion: Acidogenic Digestion*

In anaerobic digestion, four stages of fermentation are identified by the nature of the conversion products. The first stage (hydrolysis) converts the long polymers, both protein and carbohydrate, into smaller molecules that can be absorbed and digested by microbes. The second stage (acidogenesis) generates a mixture of organic acids, such as acetic, propionic, and butyric acids, along with hydrogen and CO_2. The third stage (acetogenesis) converts second stage products into acetic acid. The fourth stage (methanogenesis) generates methane. The methanogenic organisms responsible for the fourth stage are relatively sensitive to environmental conditions and will often stop growing as a result of process upsets, such as a change of pH or chemical environment. This results in a "Sour" digester which accumulates organic acids from the second and third stages and stops growing as pH is lowered. The goal of acidogenic digestion is to operate continuously in the "Sour" mode by inhibiting the methanogens while neutralizing the acids to enable continued growth of the microbial ecosystem. It is believed that the limit of acid accumulation is usually reached when the salt concentration, resulting from neutralization of acids, becomes too high for the cultures. Accordingly, microbial cultures derived from salty environments have been found to be helpful for improving yields and conversion rates in acidogenic digestion (Agbogbo and Holtzapple 2007).

9.5.4 Additional Conversion Methods or Upgrading Requirements

Several chemical pathways are possible for converting organic acids to fuels or other chemical products. Carboxylic acids, particularly short carbon chain acids such as acetic or propionic acids have relatively high oxygen content for use as a fuel. Thus, after fermentation, further conversions are required to reduce the oxygen content of these molecules. In general two approaches have been proposed: adding hydrogen to remove oxygen as water, or sacrificing carbon atoms to remove oxygen as CO_2. The first approach can be done by esterifying the carboxylic acids with an alcohol, and then hydrogenolyzing the ester to two alcohols. One alcohol can be recycled to repeat the process, the other serves as an alcohol fuel product. Through this process two hydrogen molecules are consumed, one water molecule is released, and carbon is retained. Thus, acetic acid would yield ethanol, propionic acid yields propanol, and so on. This process consumes two hydrogen molecules per alcohol produced. Equations (9.1)–(9.3) represent the generalized chemistry:

$$\mathrm{R}_1COOH + \mathrm{R}_2COH \rightarrow \mathrm{R}_1COOCR_2 + \mathrm{H}_2\mathrm{O} \tag{9.1}$$

$$\underline{\mathrm{R}_1COOCR_2 + 2\mathrm{H}_2 \rightarrow \mathrm{R}_1COH + \mathrm{R}_2COH} \tag{9.2}$$

$$\mathrm{R}_1COOH + 2\mathrm{H}_2 \rightarrow \mathrm{R}_1COH + \mathrm{H}_2\mathrm{O} \tag{9.3}$$

A second method to reduce oxygen involves neutralizing two acid molecules with a divalent base, such as lime (CaO), producing a carboxylate salt and water. The organic salt can be thermally decomposed into a ketone and alkali carbonate, such as $CaCO_3$, which has the effect of eliminating oxygen from the starting materials as CO_2. The ketone can then be easily hydrogenated to an alcohol. Figure 9.1 shows a GC-MS analysis of some of the compounds generated by the ketonization reaction.

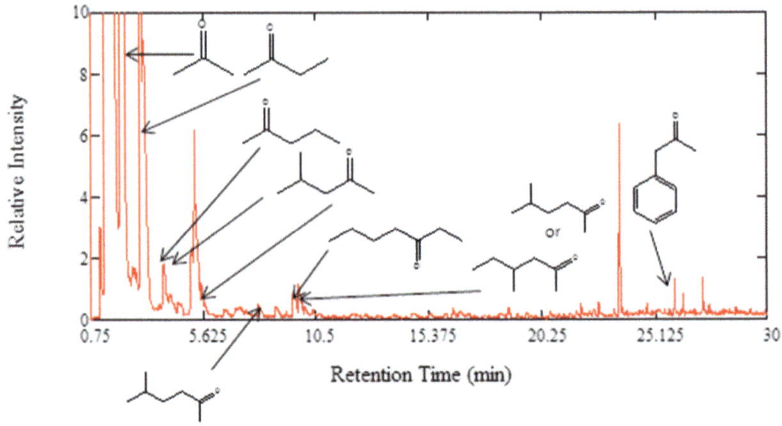

Fig. 9.1 GC-MS analysis of products present in the ketone product derived from fermentation acids

Producing alcohols via the ketonization pathway consumes less hydrogen than the esterification pathway, but reduces the carbon yield in the final fuel product. Accordingly, the ketonization route yields less fuel, but consumes less hydrogen, and the net chemical energy balance between the upgrading methods is similar. This pathway produces longer carbon chain alcohols than the esterification route, which makes the product fuel more similar to hydrocarbon fuels. The generalized chemistry is shown in (9.4)–(9.7):

$$\mathrm{R_1}COOH + \mathrm{R_2}COOH + CaO \rightarrow \mathrm{R_1}COOCaOOCR_2 + \mathrm{H_2O} \tag{9.4}$$

$$\mathrm{R_1}COOCaOOCR_2 \rightarrow \mathrm{R_1}COR_2 \underline{+CaCO_3} \tag{9.5}$$

$$\underline{2\mathrm{R_1}COR_2 + \mathrm{H_2} \rightarrow 2\mathrm{R_1}COHR_2} \tag{9.6}$$

$$\mathrm{R_1}COOH + \mathrm{R_2}COOH + CaO + \mathrm{H_2} \rightarrow \mathrm{R_1}COHR_2 + CaCO_3 + \mathrm{H_2O} \tag{9.7}$$

Both of these pathways produce mixtures of alcohols that can be used as a liquid fuel for spark ignited engines.

9.5.5 Purification

The alcohol products from the esterification and ketonization pathways emerge from the chemical conversions relatively pure. In the esterification process, distillation steps are used to separate the esters from fermentation broth, and then again to separate the final alcohols into product (short chain length), and recycle (long chain length) alcohols. Thus, the final alcohol product is relatively well separated from its parent materials. In the ketonization route, the pyrolytic decomposition of carboxylate salts to ketones will also decompose residual lignin and other organic materials. Thus, the ketone intermediate product is a very complex mixture and is often miscible with water (Chen et al. 2011; Karunarathne et al. 2010). Purification of the product alcohols and oils is most effectively accomplished by distillation of the final hydrogenated products.

9.5.6 Description of the Product

Typically the esterification route would produce a mixture containing a high proportion of ethanol (from acetate) along with other longer chain n-alcohols, while the ketonization route would produce a mixture rich in isopropanol (from acetate) and higher chain length isoalcohols. Depending on the separation methods used to isolate the carboxylate salts from the fermentation environment, the ketonization pathway could also include pyrolysis products derived from lignin or other organic materials in the fermentation mixture. These mixtures of alcohols can also be further hydrotreated to produce completely deoxygenated hydrocarbons for use in production of heavier fuel applications such as jet, diesel, or fuel oil applications (Taco-Vasquez and Holtzapple 2013).

9.5.7 Application to Forest Resources

In theory, acidogenic digestion should be able to decompose any material that is digestible in anaerobic environments. Thus, typically all carbohydrates, proteins, sugar acids, and the like can be consumed. Lignin however remains largely recalcitrant in the absence of oxygen and so is not expected to convert under acidogenic digestion. Generally, well-pretreated materials will digest more quickly, and hardwood typically pretreats more easily than softwood. Anaerobic digestion can be successfully implemented at virtually any scale, as the equipment is low cost and the operating conditions are mild. Acidogenic digestion then could be carried out at smaller scale in the field. The upgrading processes, particularly any hydrogen treatment, would best be done at a centralized, larger capacity facility. Thus it is envisioned that a hub and spoke system, akin to the deployment of mobile chippers delivering chips to a central shipping or processing terminal, may be a viable configuration for implementation in forest environments.

9.6 Thermal Deoxygenation (TDO)

The ketonization pathway described above in the carboxylate platform process can also be applied to carboxylic acids derived from biomass through chemical, rather than fermentative means. Thus, organic acids derived from biomass through acid-catalyzed hydrolysis and dehydration can be neutralized to carboxylate salts, ketonized to mixed ketones and hydrocarbons and then hydrotreated to produce a final fuel product.

9.6.1 Feedstock Conversion

This technology platform produces hydrocarbon oils and chemical byproducts from the carbohydrate fraction of the biomass. The carbohydrates are hydrolyzed and dehydrated to a mixture of organic acids (primarily levulinic and formic acids) and furfural. The organic acids are then neutralized and the resultant salts are pyrolyzed to produce a hydrocarbon oil which has characteristics that are similar to petroleum crude oils (Schwartz et al. 2010). The pyrolysis of the mixed acid salts is referred to as thermal deoxygenation, or TDO. Since pentose sugars convert to furfural rather than organic acids, the oil is ultimately produced from the hexose carbohydrates in the feed, and therefore this platform is most effective for feedstocks having significant cellulose content such as softwood or waste paper. Furfural can either be sold as a co-product, or converted to more oil using hydrogen and an additional processing step.

9.6.2 Hydrolysis and Dehydration

In the TDO process, the derivation of levulinic and formic acids from hexan polymers is accomplished by acid catalyzed hydrolysis and dehydration. Typically sulfuric acid is used as the catalyst and the breakdown of hexan to levulinic and formic acid follows well known pathways (9.8)

$$\begin{aligned} &\textit{Hexan polymers} \rightarrow \textit{hexose monomers} \rightarrow 5-OH\ \textit{methylfurfural} \\ &\rightarrow \textit{levulinic acid} + \textit{formic acid} \end{aligned} \tag{9.8}$$

9.6.3 Primary Conversion: Thermal Deoxygenation

In the same manner as ketonization is applied to fermentation-derived organic acids, neutralization and thermal decomposition of levulinic and formic acids will yield energy-densified liquid products. In addition, however, the nature of levulinic and formic acids lends advantageous features to the thermal decomposition process. As is done in formate-assisted pyrolysis, the inclusion of formate alkali salts in the pyrolysis feedstock generates active hydrogen in the reaction, which tends to

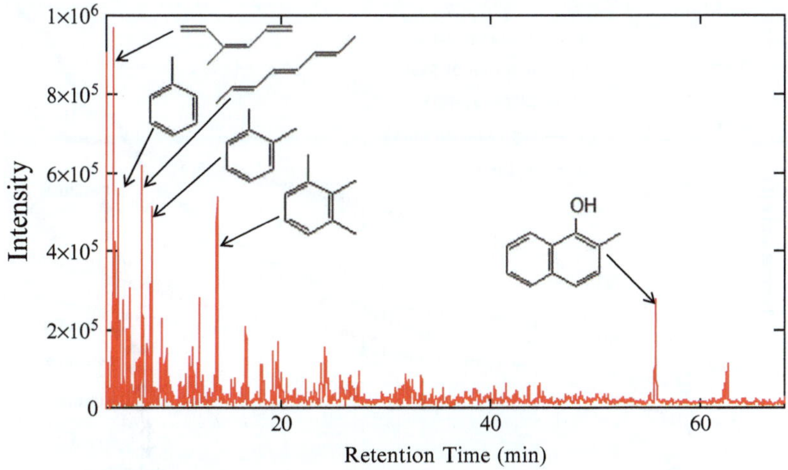

Fig. 9.2 GC-MS analysis of TDO oil derived from levulinic and formic acids

deoxygenate the pyrolysis products. In addition, the bifunctional nature of levulinic acid, having both the carboxyl group and an alcohol group, tends to lead to cyclic and linked carbon chains. This increases the molecular weight of the fuel molecules and brings the distillation range of the product oil further into the diesel and jet fuel range (Wheeler et al. 2013). It has been found that equimolar ratios of levulinic and formic acids give the best results in terms of deoxygenation and energy density (Case et al. 2012). Conveniently, the stoichiometry of the hydrolysis and dehydration reactions shown in (9.8) yields an equimolar ratio of these two acids. Figure 9.2 shows GC-MS results of an analysis of TDO oil. Note the near complete lack of oxygenated molecules. Figure 9.3 shows the boiling point distribution of a TDO oil sample ("levulene") and comparable crude petroleums and #2 diesel oil. There is about an 80 % overlap between the levulene and #2 diesel oil.

9.6.4 Additional Conversion Methods or Upgrading Requirements

Whole TDO oils, and distillate fractions derived from the oils, have been shown to function in diesel and spark ignition engines, respectively. The properties of these fuels can be further enhanced by hydrotreatment over a nickel catalyst. This saturates many of the aromatic and double bonded structures and results in fuels that are more chemically similar to petroleum-derived fuels. The application of nickel as a catalyst,

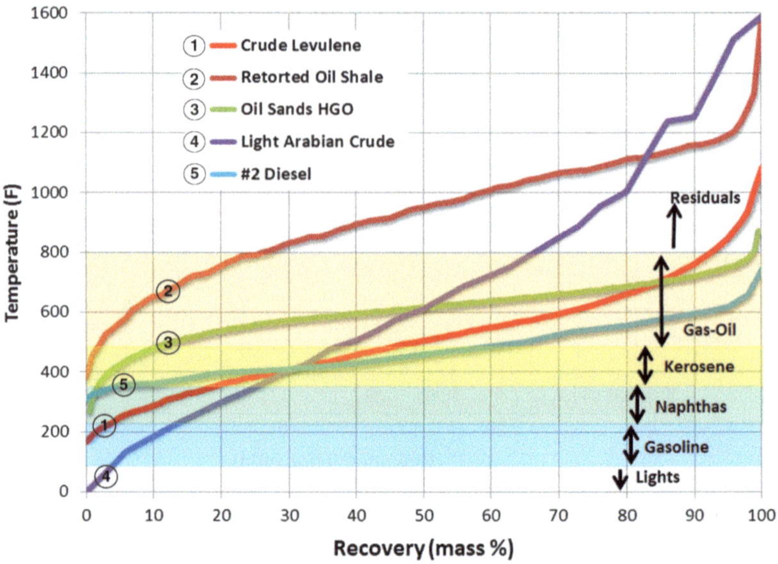

Fig. 9.3 Boiling point distribution of TDO oil ("levulene") and comparable petroleum crudes and #2 diesel fuel

which is a robust and inexpensive catalytic material, contributes favorably to the economics of the hydrotreatment process and offers a decided advantage over other pyrolysis systems requiring much more expensive and highly engineered catalysts.

9.6.5 Purification

Given sufficient formate in the acid mixture, TDO oils will separate from water and can thus be easily purified of water soluble contaminants. The resulting hydrophobic oil can be distilled in the same manner as petroleum-derived oils.

9.6.6 Description of the Product

TDO oils, even without hydrotreating, have advantageous fuel properties. The oils are stable and do not polymerize over time, as most pyrolysis oils are want to do. The oils contain very low amounts of oxygen and have low total acid numbers (TAN). Fuel tests on whole TDO oil blended 50/50 with diesel oil showed that the engines could run on this simple blend, though flame temperature was higher and flame propagation faster than normal diesel. A distillate cut from the whole oil taken across the gasoline boiling range gave a green tinted liquid that was capable of being combusted in a gasoline engine when blended with 2 % lubricating agent. Thus, there may be fuel applications for whole TDO oil or simple distillate cuts as blending agents in petroleum sources fuels. Hydrotreating the oils lightens the color of the liquid, in some cases approaching water-like clarity and colorlessness, and the chemical characteristics of the fuel are surprisingly close to those of gasoline and jet fuels.

9.6.7 *Application to Forest Resources*

Production of levulinic acid has been demonstrated at a 1 t/day throughput, and larger plants have been designed and run. Scale-up of this technology to accommodate biofuel scale would require extensive engineering, but is theoretically possible. In the near term, it is likely that levulinic acid, and possibly formic acid, would retain more value as chemical products than fuel intermediates, although selling into these markets would require additional separation effort and thus may not necessarily present an insurmountable obstacle to the development of biofuel production.

Unlike other conversion platforms, TDO may work better with softwoods, since they have more abundant 6-carbon sugars available to make levulinic acid than do hardwoods. Thus, the relative simplicity of this technology and the suitability to softwood feedstocks may make this an attractive option for biofuel production in some areas.

9.7 Summary and Comparison Table

Many pathways exist for conversion of lignocellulose to liquid transportation fuels. To date, large commercial production of liquid biofuels has been established only for fuels derived from food crop feedstocks. Most recently (DOE 2013), commercial ventures have just started to produce biofuels at large scale from lignocellulose feedstocks. In the USA, most early adopting projects are focusing on corn stover as a feedstock. Comparable technologies for ethanol production could be developed using hardwood feedstocks, and at this time softwoods are considered more risky because of their more challenging pretreatment and hydrolysis characteristics. Given that the scale of biomass availability is at least an order of magnitude lower than that realized in the petroleum industry, conversion processes that can be scaled down to match biomass availability are of critical interest. Leading technologies in this regard include fermentation and pyrolysis approaches, and possibly some innovative combinations of these two. Table 9.2 summarizes the characteristics of the technologies discussed in this chapter.

Table 9.2 Summary of biofuel conversion methods and application to wood feedstocks

Platform	Fuel product	Demonstrated scale (ton/day dry biomass)	Anticipated scale (ton/day dry wood)	Preferred wood feedstock	H_2 required	Process complexity	Current process maturity
Sugar platform alcohols	Ethanol	2,500 on corn	>2,000	Hardwood	None	Moderate	High
Sugar platform alcohols	Butanol	250 on starch or sugar	>2,000	Hardwood	None	High	Moderate
Sugar platform HC's	Bio-diesel	250 on veg. oil	>2,000	Hardwood	None	Moderate	Low
Sugar platform HC's	HC's[a]	Lab scale	>2,000	Hardwood	None	Moderate	Experimental
Pyrolysis + upgrading	HC's[a]	50	>500 in field, >2,000 upgrading	Low lignin	High	Moderate	Moderate
Formate-assisted pyrolysis	HC's[a]	Lab scale	>500 in field, >2,000 upgrading	All wood	High	Moderate	Low
Carboxylate platform (MixAlco)	Mixed alcohols (ketone route)	10 on MSW feedstock	>500 in field, >2,000 upgrading	Hardwood	Low	Low	Moderate
Carboxylate platform (MixAlco)	HC's[a]	Lab scale	>500 in field, >2,000 upgrading	Hardwood	Moderate	Moderate	Moderate
TDO	TDO oil (low H content)	1 for hydrolysis, labscale for TDO	>100 in field, >500 upgrading	Softwood	None	Low	Low
TDO	HC's[a] (high H content)	Labscale	>100 in field, >500 upgrading	Softwood	Moderate	Moderate	Moderate

[a]HC's = hydrocarbons chemically similar to petroleum-derived products

References

Agbogbo FK, Holtzapple MT (2007) Fixed-bed fermentation of rice straw and chicken manure using a mixed culture of marine mesophilic microorganisms. Biores Technol 98(8):1586–1595

Bridgwater AV (2000) Fast pyrolysis processes for biomass. Renew Sustain Energy Rev 4(1):1–73

Case PA, van Heiningen ARP, Wheeler MC (2012) Liquid hydrocarbon fuels from cellulosic feedstocks via thermal deoxygenation of levulinic acid and formic acid salt mixtures. Green Chem 14(1):85–89

Chen X, Wheeler MC, van Walsum GP (2011) Bioconversion of brown seaweed into mixed ketones through the carboxylate platform processing. Conference proceedings of the AIChE annual meeting, Minneapolis, MN, 17–20 Oct 2011

DOE (2013) Presentations made at the 2013 DOE Bioenergy Peer Review, Washington DC, 20–23 May

Elliott DC (2007) Historical developments in hydroprocessing bio-oils. Energy Fuels 21(3):1792–1815

Furimsky E (1983) Chemistry of catalytic hydrodeoxygenation. Catalysis Rev 25(3):421–458

Holtzapple MT, Davison RR, Kyle Ross M, Aldrett-Lee S, Nagwani M, Lee CM, Lee C, Adelson S, Kaar W, Gaskin D, Shirage H, Chang N-S, Chang VS, Loescher ME (1999) Biomass conversion to mixed alcohol fuels using the MixAlco process. Appl Biochem Biotechnol 77–79:609–630

USEIA (Energy Information Administration) (2013) Number and capacity of petroleum refineries. http://www.eia.gov/dnav/pet/pet_pnp_cap1_dcu_nus_a.htm. Accessed 15 Aug 2013

Karunarathne SA, Wheeler MC, van Walsum GP (2010) Production of carboxylic acids from acidogenic fermentation of Algefiber® (sea weed sludge) using a mixed culture of marine microorganisms. Conference proceedings of the AIChE annual meeting, Salt Lake City, UT, 7–12 Nov 2010

Lynd LR, Wyman CE, Gerngross TU (1999) Biocommodity engineering. Biotechnol Prog 15(5):777–793

Mukkamala S, Wheeler MC, van Heiningen ARP, Desisto WJ (2012) Formate-assisted fast pyrolysis of lignin. Energy Fuels 26(2):1380–1384

Schwartz TJ, van Heiningen ARP, Wheeler MC (2010) Energy densification of levulinic acid by thermal deoxygenation. Green Chem 2010(12):1353–1356

Strahan GD, Mullen CA, Boateng AA (2011) Characterizing biomass fast pyrolysis oils by ^{13}C NMR and chemometric analysis. Energy Fuels 25(11):5452–5461

Taco-Vasquez S, Holtzapple MT (2013) Conversion of isopropanol and mixed alcohols to hydrocarbons using HZSM-5 catalyst in the MixAlco process. AIChE J 59(7):2549–2557

Wheeler MC, Van Walsum GP, Schwartz TJ, van Heiningen ARP (2013) Energy densification of biomass-derived organic acids. U.S. patent 8,362,306

Zhang H, Vispute TP, Xiao R, Huber GW (2011) Catalytic conversion of biomass-derived feedstocks into olefins and aromatics with ZSM-5: the hydrogen to carbon effective ratio. Energy Environ Sci 4(6):2297–2307

Chapter 10
Emissions and Health Effects of Wood Biomass Combustion

Polly Hoppin and Molly Jacobs

10.1 Introduction

Many industrial, commercial, and institutional (ICI) wood combustion units emit higher concentrations of hazardous pollutants—including fine particulate matter, volatile organic compounds, and carbon monoxide—than do comparably sized units burning other fuels, such as oil and natural gas (Associates and Northeast States for Coordinated Air Use Management 2009; NARSTO 2010; Hoppin and Jacobs 2012; Booth 2012a). Yet the potential risks to human health from emissions and from the production process of wood biomass energy have not been systematically addressed in policies and programs promoting wood combustion as a renewable energy strategy. Programs providing subsidies or other incentives for biomass applications focus their communications on the societal benefits of a shift to wood as fuel, as discussed in other chapters. At the project level, health concerns have surfaced in communities where large-scale biomass facilities are proposed, but are largely absent in discussions about smaller scale installations in apartments and schools, municipal buildings, health care facilities, and commercial settings. Particularly in the context of the proliferation of wood biomass as a source of heat and power in communities with other sources of pollution and with susceptible populations, it is important to understand the health hazards of wood combustion and opportunities for reducing risks.

This chapter provides an overview of what is known and not known about emissions from wood combustion, with a particular focus on the nonresidential scale. In addition, it reviews the state of the evidence on health effects associated with exposure to emissions from wood combustion, information that is relevant to

P. Hoppin, Sc.D. (✉) • M. Jacobs, M.P.H.
Lowell Center for Sustainable Production, School of Health Sciences, University of Massachusetts, Wannalancit Building, 600 North Suffolk Street, Lowell, MA 01854, USA
e-mail: Polly_Hoppin@uml.edu; molly_jacobs@uml.edu

M. Jacobson and D. Ciolkosz (eds.), *Wood-Based Energy in the Northern Forests*, DOI 10.1007/978-1-4614-9478-2_10, decisions about wood biomass applications at any scale. Finally, the chapter describes policy approaches available for protecting public health from pollution and concludes with solutions: recommendations for changes in technologies, policy and programs to better protect public health. While the discussion focuses on the Northeast region of the USA, it has broader implications, providing a framework for incorporating health in renewable energy decision-making across the country.

10.2 Pollutants Emitted During Wood Combustion: Health Hazards and Exposure Considerations

The US Environmental Protection Agency (EPA) has identified over 100 chemical pollutants in the mixture of wood combustion emissions. Table 10.1 summarizes known health hazards of these pollutants.

10.2.1 Pollutants of Particular Concern: Particulate Matter, Carbon Monoxide, Polycyclic Aromatic Hydrocarbons, and Heavy Metals

Pollutants currently understood to be of greatest health concern for wood combustion systems include: particulate matter (especially fine particulates less than 2.5 μm in size($PM_{2.5}$)), carbon monoxide (CO), polycyclic aromatic hydrocarbons (PAHs), and if the wood is treated or contaminated, heavy metals.

Particulate Matter. Particulate pollution from wood biomass combustion is well characterized and can be of concern for public health. Particulate pollution is a complex mixture of diverse substances that include solid particles, liquid particles or droplets, but not gaseous compounds. The mixture includes toxic and irritant acids, organic chemicals, metals, and soil or dust particles.

The size of the particles is relevant to their potential to impact health. Most of the particulate matter in wood biomass emissions is invisible to the naked eye, as the particles are smaller than 10 μm (PM_{10}) and dominated (on a number of particles basis) by ultrafine particles with an aerodynamic diameter of less than 1 μm (Sippula et al. 2009). (For size reference, the diameter of a human hair is ~50–70 μm and beach sand is ~90 μm in diameter.) Ultrafine particles have a short life in the atmosphere (minutes to hours) and rapidly grow (through coagulation and/or condensation) to form larger complex aggregates, but typically stabilize at a size less than 2.5 μm ($PM_{2.5}$) (Pope and Dockery 2006). Both ultrafine and $PM_{2.5}$ particles tend to deposit in the deep lung (alveolar regions), where they penetrate the blood stream and can have systemic biological effects. $PM_{2.5}$ from wood biomass combustion can remain suspended in ambient air for longer periods of time, can be transported over long distances, and can penetrate more readily into indoor

Table 10.1 Classes and examples of chemical hazards in emissions from wood-fired combustion units (Naeher et al. 2007; Washington State Department of Ecology 2003; Association of Occupational and Environmental Clinics (AOEC) 2010; International Agency for Research on Cancer 2013; Agency for Toxics Substances Disease Registry 2010)

Chemical class	Primary examples	Known health hazard
Criteria pollutants	Carbon monoxide	Cardiac
	Lead and compounds	Carcinogen[a], cardiac, reproductive/developmental toxicant
	Nitrogen oxides	Asthmagen[b], immune toxicant, other respiratory toxicant,
	$PM_{2.5}$	Asthmagen[b], cardiac, other respiratory toxicant
	PM_{10}	Asthmagen[b], cardiac, other respiratory toxicant
	Sulfur dioxide	Asthma (exacerbation), other respiratory toxicant
Inorganic acids	Hydrochloric acid	Corrosive, respiratory irritant
	Sulfuric acid	Corrosive, respiratory irritant
Metals (20+ potential compounds)	Arsenic	Carcinogen[c], cardiac, other respiratory toxicant, reproductive/developmental toxicant
	Cadmium	Carcinogen[c], endocrine disruption (osteoporosis), respiratory toxicant, reproductive/developmental toxicant
	Chromium and compounds	Asthmagen[b], carcinogen[c], other respiratory toxicant, reproductive/developmental toxicant
	Mercury	Reproductive/developmental toxicant
	Nickel and compounds	Asthmagen[b], carcinogen[c] (nickel compounds), carcinogen[a] (nickel), immune toxicant, other respiratory toxicant
Polyaromatic hydrocarbons (PAHs) (20+ potential compounds)	Benzo(a)pyrene	Carcinogen[c], immune toxicant
	Indeno(1,2,3,cd)pyrene	Carcinogen[a]
	Naphthalene	Asthmagen[b], carcinogen[a]
Volatile organic compounds (VOCs) (20+ potential compounds)	Acetaldehyde	Asthmagen[b], carcinogen[a], respiratory irritant
	Acrolein	Respiratory irritant
	Formaldehyde	Asthmagen[b], carcinogen[c], respiratory irritant
	Benzene	Carcinogen[c], cardiac, reproductive/developmental toxicant, immune toxicant, neurotoxicant
	Toluene	Neurotoxicant, respiratory irritant
	1,3 Butadiene	Carcinogenic[c], respiratory irritant
	n-Hexane	Neurotoxicant
Dioxins, furans and other "dioxin-like" compounds	2,3,7.8 tetrachlorodibenzo-p-dioxin; 2,3,7.8 tetrachlorodibenzo-p-furan	Carcinogen[c], cardiac, endocrine disruption (type II diabetes, thyroid disorders), immune toxicant, reproductive/developmental toxicant

[a]Classified as a "probable" or "possible" human carcinogen by IARC or "reasonably anticipated" to be a human carcinogen the US NTP, respectively

[b]Classified as an agent that can cause asthma de novo by the Association of Occupational and Environmental Clinics (also associated with triggering subsequent attacks)

[c]Classified as "known" to be a human carcinogen by the International Agency for Research on Cancer(IARC) and the US National Toxicology Program (IARC)

environments as compared to larger, coarser particles (Wilson and Suh 1997). Consequently, fine particulate emissions into the outdoor ambient air from institutional, commercial, industrial, and electricity-generating wood biomass facilities can potentially pose an indoor as well as outdoor air quality hazard (Allen et al. 2003; Larson et al. 2004).

Carbon monoxide. Carbon monoxide (CO) emissions are products of incomplete and inefficient combustion. Good combustion practices in wood biomass units (reviewed further below) are directly associated with lower levels of CO.

PAHs. Though quantities of PAHs emitted from wood combustion are generally lower than from coal, some studies have found that wood combustion emits more carcinogenic or genotoxic PAHs per unit of fuel energy than several types of coal emit (Oanh et al. 1999; Truesdale and Cleland 1982). Additional studies comparing small institutional and commercial-sized wood-fired boilers to equivalent oil boilers suggest that PAH levels can be higher for wood-fired boilers, especially for older boilers and for boilers with high CO emissions—an indicator of incomplete combustion (Johansson et al. 2004).

Heavy metals. Trace levels of heavy metals such as arsenic, cadmium, and nickel are generally found in the air emissions of wood-fired combustion units because the metals occur naturally in many types of wood (Washington State Department of Ecology 2003; Demirbas 2008). Metal emissions can be highly variable depending on the type of wood or biomass being burned, whether the fuel includes bark, whether the fuel is contaminated with other debris, and whether other fuel sources are used in combination with wood (Demirbas 2008; Chandrasekaran et al. 2012). While metals are not a significant fraction of particulate pollution, one recent study of emissions from institutional and commercial wood biomass combustion units (2.8–16.4 MW) found that heavy metals and trace elements, which occur naturally in wood fuel, showed a tendency to concentrate in fine particles (Sippula et al. 2009).[1]

10.2.1.1 Burning Construction and Demolition Debris

Among Northeast states, facilities in Maine and one facility in Connecticut currently have permits to burn C&D wood debris. Emissions from the combustion of C&D wood debris are determined by what compounds are contained in the wood debris fuel, and can include heavy metals such as lead and arsenic, dioxins/furans, and volatile organic compounds, among other pollutants (Northeast States for Coordinated Air Use Management 2006; Maine Department of Environmental Protection 2007). Regulations in Maine restrict the percentage of C&D debris in the fuel supply to 50 % (Maine Department of Environmental Protection 2007). Maine has adopted fuel quality standards that set limits on the quantities of treated wood and non-wood materials that may be present prior to blending with C&D wood, and

[1] Samples in this study were taken after passing through emission controls.

has set limits on the concentration of several chemical contaminants after blending. Despite the fuel quality standards, a state review of C&D wood fuel samples from 12 facilities identified five facilities that failed to meet the arsenic standard and two that failed to meet the treated wood standard, though all were below the lead standard (Maine Department of Environmental Protection 2007). These findings are consistent with studies demonstrating that simple visual sorting—the method used by processing facilities to identify and remove unpermitted material—is not fully effective for distinguishing treated wood from untreated wood (Solo-Gabriele et al. 1999; Holton 2001).

Assessments of C&D wood debris emissions in the Northeast are based primarily on assessments at three facilities in Maine (Northeast States for Coordinated Air Use Management 2006). These assessments suggest that heavy metal and dioxin emissions are higher from the burning of C&D wood debris compared to combustion of non-C&D wood (Maine Department of Environmental Protection 2007; Humphrey 2005).

10.2.2 Variability in Emissions

The composition of pollutants and concentrations emitted from wood biomass combustion units vary widely. The factors that drive this variability provide opportunities for reducing public health risks. These include the following:

1. Type of wood fuel
2. Moisture content of the wood
3. Type of combustion technology
4. Operation characteristics of the combustion unit
5. Type of emission control technology used

Type of wood fuel. The chemical and physical composition of different wood biomass fuels can vary substantially, resulting in different types and concentrations of substances emitted during combustion. For example, the higher the ash content (mineral content), the higher the particulate and toxic emissions, including acid gas emissions and heavy metals among others (Demirbas 2008; Beauchemin and Tampier 2008). Bark contains more ash than wood because of significantly higher mineral content (Demirbas 2008). Lastly, oversized pieces are difficult to distribute evenly in the boiler and tend to burn slowly resulting in higher levels of pollutants that are products of incomplete combustion, such as CO and PAHs. As a consequence, burning wood pellets tends to emit lower levels of pollutants than does burning wood chips, which are less polluting than wood logs. However, a recent study found evidence of heavy metal contamination (including arsenic and lead, among other metals) of commercially available wood pellets (Chandrasekaran et al. 2012).

Moisture content of the wood. When wood is burned, the water in the fuel must be driven off before the first stage of combustion can occur. Wood with high moisture content can result in a lower than optimal temperature in the combustion chamber. As a consequence, incomplete combustion of the fuel occurs and pollutants such as CO, fine particulates, and PAHs can be emitted. In general, the higher the moisture content of the wood, the more incomplete the combustion and the higher the levels of pollutant emissions (Bignal et al. 2008; Atkins et al. 2010). Conversely however, NOx emissions from wet bark are typically lower (approximately one-half) compared to NOx emissions from dry wood (Environmental Protection 1995a).

Type of combustion technology. Technologies to convert wood fuel into energy fall into two general categories: direct-fired combustion and two-stage gasification systems. Direct-fired combustion systems work by burning wood directly, as the name implies. Stoker boilers are one of the more common direct combustion units used in the Northeast, especially for small to medium sized institutional/commercial scale systems (Hinckley and Doshi 2010). In contrast, gasification or 2-stage combustion technologies involve heating the wood biomass first in an oxygen-limited environment to release volatile gases. These gases are subsequently burned in a second chamber. Combustion in two-stage combustion units is more efficient than in direct fired combustion units, because the technology optimizes factors that drive complete combustion (temperature, time, and turbulence, see below). For example, depending on the moisture content of the wood fuel used, small stoker boilers can achieve combustion efficiencies that range from 60 to 80 %, whereas two-stage systems can achieve efficiencies greater than 80 % (Hoppin and Jacobs 2012; Urquhart and Boyce 2008). In general, the more complete the combustion, the lower the levels of emissions.

Operational characteristics of the combustion unit. Temperature, time, and turbulence are three key factors that influence combustion efficiencies and associated emissions (Musil-Schläffer et al. 2010). For example, during start-up or shutdown, the combustion chamber is ramping up to, or ramping down from the optimal steady-state combustion temperature and the optimal air-to-fuel ratio. This results in decreased combustion efficiencies and higher pollutant emissions than when the combustion chamber is operating at steady-state. Similarly, emissions are higher when boilers are operating at low versus high loads, because the combustion conditions are not optimized. Wood boilers that are oversized for the heat demand of a building can operate at either low loads for long periods of time or at rapidly changing loads–, both conditions that increase emission levels (Hoppin and Jacobs 2012).

Minimizing emission levels also requires adequate training of operators, the use of detectors that flag operational problems, and automated emission controls to optimize combustion efficiency and proper maintenance (Hinckley and Doshi 2010).

Type of emission control technology. There are four main types of "add-on" air pollution control devices for wood-fired units: mechanical collectors (cyclones, multicyclones, core separators), electrostatic precipitators (ESP), fabric filters or baghouses, and scrubbers. Emission control efficiencies among the different technologies are highly varied, especially for $PM_{2.5}$. For example, in an evaluation of a sample of small wood fired combustion units in the Northeast states, the control efficiencies for $PM_{2.5}$ ranged from only 3 % with a single cyclone up to 74 % when a

Table 10.2 US Environmental Protection Agency AP 42 emission factors (Environmental Protection 1995b)

Fuel	PM control device	Filterable PM (g/GJ)	Filterable PM 10 (g/GJ)	Filterable PM 2.5 (g/GJ)
Wet Bark/Bark and wood	No control	270	220	190
Dry wood	No control	170	160	130
Wet wood	No control	140	120	110
Bark	Cyclone	230	210	120
Bark and wet wood	Cyclone	150	140	80
Dry wood	Cyclone	130	120	70
Wet wood	Cyclone	90	90	50
All fuels	Electrolyzed gravel bed	40	32	28
All fuels	Wet scrubber	28	28	28
All fuels	Fabric filter	40	32	28
All fuels	ESP	23	17	15

multicyclone and baghouse were used (Hinckley and Doshi 2010). ESP technology is widely used in Europe and evidence suggests that control efficiencies for PM_{10} and $PM_{2.5}$ appear to be 99 % or greater (Hinckley and Doshi 2010). However, currently, ESPs are not being used on the vast majority of small boilers 0.9–2.9 MWt (3–10 million btu/hr) in the Northeast.

10.2.2.1 Comparisons of Particulate Pollution Emission Rates

Because of the range of factors above that influence wood combustion emissions, it can be misleading to generalize emissions from one unit to another and even within an individual unit; to assume, for example, that emissions from a given wood biomass combustion unit are reflected by stack tests taken during steady state (Hoppin and Jacobs 2012). Wood combustion is a dynamic process; the five factors described above interact to produce a wide variety of pollutants at varying concentrations. The influence of fuel type and emission controls on emissions is demonstrated by emission factors for particulate matter developed by EPA (Table 10.2). As shown, emission levels can vary by an order of magnitude depending simply on fuel type and the emission control technology used.

In rural states, residential use of wood fuel can be significant. As a result, the predominant source of PM from wood combustion in a given region is typically from residential use, such as wood burning stoves. For example, in Vermont, 92 % of the PM from wood combustion is estimated to be from residential sources (Snook 2012). Emission rates of PM for residential/domestic units vary depending on the technology, with pellet stoves emitting far less particulate pollution than conventional woodstoves (Table 10.3). In addition, the typical residential/domestic wood combustion technology burns far less efficiently—and is therefore more polluting on a per mass basis—than current ICI boilers. However, because the tonnage of fuel

Table 10.3 PM emission rates and estimated total yearly emissions for representative domestic institutional and commercial wood burners (Booth 2012b)

Burner Technology	$PM_{2.5}$ emission factor (Kg/tonne)	Tonnes fuel/year (dry)	$PM_{2.5}$ emissions (tonnes/year)
Conventional wood stove	33.4	2.2	0.07
Certified non-catalytic woodstove	7.3	1.9	0.02
Certified catalytic wood stove	8.1	1.9	0.02
Pellet stove	4.4	1.5	0.01
Burner Technology	PM10 emission factor (Kg/tonne)[a]	Tons fuel (green)	PM10 emissions (tons)
4 MMBtu/h (school boiler)	0.11	910	0.9
15 MMBtu/h (lumber mill)	0.12	11,400	13.5

[a]Based on using a multicyclone emission control device

burned in an ICI boiler is high compared to residential/domestic units, the resulting particulate pollution emissions on an annual basis are also higher (Table 10.3). Therefore, despite contributing a relatively low percentage of current regional loads of fine particulate pollution, a given ICI wood combustion unit can be an important contributor to poor air quality where it is located (Hoppin and Jacobs 2012). As described further below, this is especially true if facilities are located in valleys, subject to thermal inversions or other atmospheric and topographical conditions that impede air pollution dispersion.

10.2.2.2 Emissions from Advanced Wood Combustion Technologies

Advanced wood combustion technologies show promising reductions in emission levels relative to the conventional direct-fired boilers currently in wide use. These combustion systems tend to use the two-stage combustion/gasification technology.

Evaluations of emissions from three 2-stage wood biomass combustion units[2] installed in two New York institutions found that $PM_{2.5}$ emissions were lower than emissions from institutional combustion units that used the more conventional, direct-fired stoker boiler design (26–43 g/GJ compared to 112–120 g/GJ, respectively (all devices used cyclones for emission control) (Hopke and Holsen 2010).[3] Among the three 2-stage combustion units, those burning wood pellets generated lower emissions than those burning wood chips (26 g/GJ compared to 43 g/GJ). However, $PM_{2.5}$ emissions from each of the three 2-stage combustion units were higher than those of oil-fired and gas-fired combustion systems (NARSTO 2010; Hopke and Holsen 2010). Researchers estimate that if an ESP were installed on these 2-stage units, emissions of $PM_{2.5}$ would further decline to 1.1 g/GJ; levels that

[2] Unit size = 0.15–0.5 MWt heat capacity.

[3] The Environmental Protection Agency's Area Source Boiler Rule standard for boilers between 2.9 and 8.8 MWt (heat input) is 30 g/GJ.

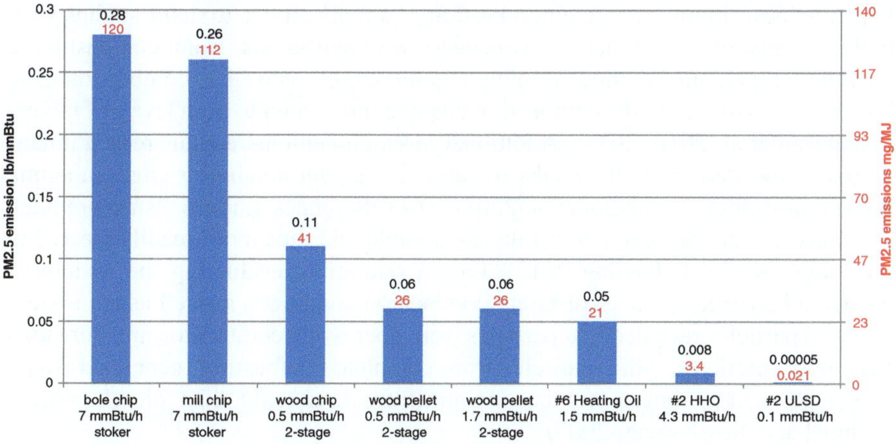

Fig. 10.1 Comparison of institutional heating technologies and fuels: PM 2.5 emissions on an energy input basis (Source: New York State Research and Development Authority (NYSERDA) Chandrasekaran et al. 2011; Rector 2010; McDonald 2009)

are lower than average emissions from both #2 fuel oil and natural gas systems, yet still higher than ultralow sulfur heating fuel oil (NARSTO 2010; Hopke and Holsen 2010). These results are shown in Fig. 10.1.

Based on a recent third-party evaluation, similar advances are also being made for larger wood biomass gasification systems for district heating and industrial uses as well (Levelton Consultants Ltd 2010).

10.2.3 Particulate Pollution: The Impact of Completeness of the Combustion on Toxicity

The completeness of the combustion process determines not only the concentration of particulate pollution, but also the physical and chemical properties of the emitted woodsmoke particles—both affect toxicity. Research suggests that woodsmoke particles can be divided into three classes based on the combustion conditions and physical and chemical properties: (1) organic carbon, (2) soot, and (3) alkali salts (Kocbach Bølling 2012). Particulate emissions from combustion units often contain a mixture of these different classes of combustion particles, yet organic carbon and soot tend to be dominant in emissions from conventional combustion systems where efficient and complete combustion is more difficult to achieve. In advanced, two-staged combustion systems, alkali salts are the primary particles emitted (Kocbach Bølling 2012).

These differences are important: research suggests that particles emitted from poor combustion systems are more toxic than those from more advanced systems which achieve more complete combustion. Animal/cellular toxicity studies show higher levels of inflammation associated with emissions from combustion at medium temperatures compared to high temperatures, as well as combustion using low levels of oxygen in the combustion chamber compared to high levels of oxygen (Danielsen et al. 2010, 2011). Additional research demonstrates increased inflammation associated with alkali salts in cell cultures, yet no similar effect in animal inhalation studies (Bellmann et al. 2008). The differences in these results are likely explained by the fact that alkali salts are soluble and thus more readily cleared by the lungs (Kocbach Bølling 2012). Lastly, one study evaluating the toxicity of woodsmoke particles under different combustion conditions as well as the toxicity of diesel particles revealed that particles from poor wood combustion are more toxic than diesel particles, while particles from complete combustion (generated from a two-stage wood biomass combustion unit) are less toxic than diesel particles (Klippel and Nussbaumer 2007).

10.2.4 Additional Factors that Affect Wood Combustion Emissions Levels: Meteorological Conditions, Topographical Conditions, and Proximity to the Source

The hazardous pollutants generated by wood combustion pose a greater or lesser threat to health depending on exposure. While the concentration of pollutants emitted from a given wood combustion source depend on factors as described above, the levels that reach a population depend primarily on three additional factors: meteorological conditions, topographical conditions, and proximity to the source. These factors, along with the activity-time patterns of an individual, will influence exposure levels.

Once emissions leave the stack of the wood biomass combustion unit, they are transported and dispersed across the landscape. Multiple factors influence what happens to the pollutants as they disperse. For example, atmospheric and metrological conditions can change the chemical composition of the original air pollutant mixture as a result of interactions/reactions with other air pollutants, water vapor, and particle surfaces (Naeher et al. 2007). Meteorological and topographical conditions can disperse pollutants and reduce the potential for human exposure to high concentrations of pollutants. On the other hand, meteorological conditions such as diurnal and multi-day atmospheric temperature inversions that occur along with low wind speed can trap and concentrate pollutants—conditions that can be common in some regions of the Northeast during winter when wood-burning for heating purposes peaks. In communities located in places with topography that tends to catch pollution such as valleys, mountainous terrain, or river basins, even a small number of wood biomass combustion units can give rise to high pollution events. For example, a recent study modeling wood smoke emissions during the winter months (primarily from residential wood burning) in New York, revealed high spikes in wood smoke concentrations (>100 μg/m^3) over the course of several minutes during the night (Miller et al. 2010). Concentrations of woodsmoke were often highest in villages and towns where the greatest concentration of people in rural communities live (Miller et al. 2010). In fact, of the seven rural counties in New York included in the study during the winter of 2009, 127,670 or 26 % of the population was exposed to high levels of PM2.5 in wood smoke (Miller et al. 2010).

Finally, proximity to combustion sources also affects pollutant concentrations. As a consequence, the percentage of pollution that is inhaled relative to the overall volume of emissions—known as the "intake fraction"—is higher the closer populations are to the source (Brauer 2012).

10.3 Health Impacts of Wood Biomass Combustion

The toxicological and epidemiological literature exploring health effects of wood combustion confirms that exposure to emissions from wood biomass combustion pose risks to public health. These include studies of people exposed to woodsmoke from residential wood combustion or wildfires, as well as controlled exposure chamber studies that examined subclinical health outcomes among healthy volunteers (Naeher et al. 2007).

10.3.1 Evidence Linking Woodsmoke Exposure with Respiratory and Cardiovascular Disease

Studies of exposure to woodsmoke demonstrate a range of harms to respiratory health in both children and adults. Health impacts include: increased health care utilization to address respiratory illnesses, aggravation of asthma, aggravation of chronic obstructive pulmonary disease, bronchitis, and decreased lung function (Naeher et al. 2007; Boman et al. 2003). Studies with more refined source apportionment methods that examine risks associated with specific woodsmoke tracers in $PM_{2.5}$ (e.g., potassium or levoglucosan) help to substantiate these findings (Ostro et al. 2009; Allen et al. 2008; Karr et al. 2009). One of these studies identified additional health risks for young children, notably an 8 % increase risk of bronchiolitis—the leading cause of hospitalizations for children under 1 year of age—among those exposed to the highest levels of woodsmoke compared to those exposed to the lowest levels (Karr et al. 2009). A recent study evaluating the impact of a wood change out program provides some data regarding the magnitude of health impacts among children that can be improved with reductions in levels of woodsmoke $PM_{2.5}$. The study found that a 5 μg/m^3 decrease in average winter time $PM_{2.5}$ was associated with a 26.7 % reduction in reported wheeze among children as well as a significant reduction in a number of respiratory infections, such as a 54.6 % reduction in bronchitis and a 25.4 % reduction in colds (Noonan et al. 2012).

In 2007, the most comprehensive review paper to date regarding the health effects of woodsmoke found no persuasive evidence that woodsmoke particles are significantly less dangerous for respiratory disease than other major categories of particulate pollution, such as vehicular emissions (Naeher et al. 2007). Indeed, in a 2003 review, Boman and colleagues had found *higher* risks for respiratory outcomes from woodsmoke in studies where residential wood combustion was compared to other sources of particulate pollution (Boman et al. 2003).

Evidence linking woodsmoke exposure to increased cardiovascular outcomes is more limited than evidence linking woodsmoke exposure with respiratory disease (Naeher et al. 2007). Though the research is relatively sparse, the studies that have been conducted demonstrate mixed findings for associations between woodsmoke and cardiovascular effects such as mortality and cardiac arrest in communities affected by wildfires (Dennekamp et al. 2011; Henderson et al. 2009; Delfino et al. 2009). However, there is evidence among wildland firefighters of systemic inflammation—an early marker of cardiovascular disease—as well as evidence of reductions in subclinical indicators of cardiovascular morbidity among healthy individuals as exposure to levels of woodsmoke $PM_{2.5}$ are reduced (Swiston et al. 2008; Allen et al. 2011). Yet in controlled exposure studies of volunteers, evidence associating systemic inflammation with exposure to woodsmoke was mixed (Sallsten et al. 2006; Barregard et al. 2006, 2008).

Despite the need for more research on the topic of cardiovascular effects associated with woodsmoke exposure, there is persuasive evidence of associations between exposure to particulate pollution (predominantly from mobile sources) and cardiovascular outcomes. As summarized by the 2010 scientific statement of the American Heart Association (AHA) (Brook et al. 2010) there is clear evidence that elevations in exposure to fine particulates is strongly associated with:

- Increased cardiovascular mortality from short-term exposure periods in ambient air
- Increased hospitalizations for cardiac conditions after short-term periods
- Increased ischemic heart disease events as a result of both short and long-term periods
- Altered heart rate (typically reduced variability) after short-term periods

In addition to these health outcomes, there is evidence that short-term exposure to $PM_{2.5}$ is associated with stroke and heart failure as well as subclinical cardiovascular conditions such as systemic inflammation, increased blood coagulation, increased blood pressure, and increased vascular dysfunction (Brook et al. 2010).

The US EPA lowered the annual PM2.5 standard to 12 $\mu g/m^3$ in 2013, but studies have observed premature mortality at much lower levels (Crouse et al. 2012; Krewski et al. 2009). Numerous epidemiological studies examining the relationship between increase in $PM_{2.5}$ pollution and the most serious adverse health outcomes such as premature death or hospitalizations associated with heart or pulmonary conditions have not been able to identify a level of "no risk" (i.e., a no-threshold model) (Pope and Dockery 2006; Brook et al. 2010). This implies that health protection is expected to improve with reductions in air pollution at any level.

10.3.2 Evidence Linking Woodsmoke and Pollutants in Woodsmoke with Cancer

Cancer is also a health outcome of concern, given the broad array of known mutagens and carcinogens in the woodsmoke mixture, including PAHs, VOCs such as benzene and formaldehyde, and heavy metals, including arsenic and cadmium among others. In 2010, the International Agency for Research on Cancer published its evaluation of the carcinogenicity of biomass combustion. Their review, based primarily on studies of indoor household wood burning, was that wood biomass emissions are "probably carcinogenic" to humans (International Agency for Research on Cancer 2010). Brown and colleagues recently estimated that cancer risk for people exposed to PAHs in residential outdoor wood boiler smoke was 7–100-fold higher than for people not exposed (Brown et al. 2007).

10.3.3 Evidence Linking Woodsmoke and Pollutants in Woodsmoke with Other Health Outcomes

Additional notable health outcomes that have been linked to woodsmoke exposure include an increased risk of otitis media (middle ear infections). Recent research in Vancouver British Columbia identified a 30% increased risk of otitis media among infants and children exposed to heavy woodsmoke pollution compared to those exposed to the lowest levels (Karr et al. 2009; MacIntyre et al. 2011). Similar results have also been observed in a case-control study in rural New York among children whose families used wood burning stoves (Daigler et al. 1991).

10.3.4 Additional Concerns for the Health of Workers and Communities: Upstream Hazards from the Production, Transportation, and Storage of Wood Fuel

Risks from exposure to woodsmoke are not limited to stack emissions from wood-burning units. Workers who make wood chips and pellets are at risk of accidental injury as well as health impacts associated with wood dust, molds and endotoxins. Occupational exposure to wood dust is associated with a broad range of respiratory effects, including rhinitis, sinusitis, nasal obstruction, asthma, and chronic bronchitis as well as dermatitis (Demers and Teschke 1998). Molds growing on wood, especially wood bark, have been associated with acute respiratory illness and hypersensitivity pneumonitis among workers exposed to wood dust and wood chips (van Assendelft et al. 1985; Asmussen 1986). Research on workers at wood biomass power plants has demonstrated that those most highly exposed to endotoxin levels in wood dust and wood dust itself are at increased risk of asthma and other work-related respiratory symptoms (Schlunssen et al. 2011). Recent incidents involving explosions at wood pellet plants implicate combustible dust as an explosion hazard (Press Release 2011a, b; Inspection: 313204299—Creative Biomass Inc 2011). OSHA has fined several pellet manufacturing plants as a result of such incidents (Press Release 2011a, b; Inspection: 313204299—Creative Biomass Inc 2011).

Another important exposure of concern is diesel exhaust from the heavy-duty trucks used to transport wood fuel. As is the case with woodsmoke itself, exhaust from diesel engines contains $PM_{2.5}$ as well as hundreds of chemicals that adsorb onto the surfaces of particles. Diesel pollutants include known mutagens and carcinogens, PAHs, and irritant and toxic gases such as VOCs and NOx, which are particularly important for local ozone formation (Health Effects Institute Air Toxics Review 2007). Facilities requiring a heavy volume of diesel truck traffic can create localized "hotspots" of air pollution along transportation corridors as well as within communities where truck engines are left idling for long periods. Several studies show that high-volume truck traffic contributes substantially to localized levels of $PM_{2.5}$ (Suvendrini Lena et al. 2002) and PAHs (Levy et al. 2003; Richmond-Bryant et al. 2009). Diesel truck traffic carrying wood fuel to large electric power-generating plants can be substantial—worst-case estimates for such a plant exceed 100 truck-loads per day (Booth 2008). Though pollutant concentrations will vary depending on meteorological conditions and type of vehicle among other factors, diesel exhaust emissions should be of concern to facility workers, truck drivers, and community residents, particularly those living near main roads.

Since 2002, 14 deaths and several nonfatal incidents have occurred as a result of the release of carbon monoxide from the storage of wood fuel, primarily wood pellets (Gauthier et al. 2012). These incidents have occurred in industrial settings, including the transport of wood fuel in cargo ships and the storage of wood pellets in silos, as well as the storage of wood pellets in private residential storeroom (Gauthier et al. 2012; Svedberg et al. 2009, 2008). All victims died of carbon monoxide poisoning as a result of CO emissions from the wood fuel (Gauthier et al. 2012). Investigations launched after two of the fatalities documented that small quantities of wood pellets in confined spaces can produce life-threatening levels of CO at temperatures of 26 °C (79 °F) (Gauthier et al. 2012). Factors that influence CO emissions include the temperature of the storage room as well as the age of the pellets, with fresher pellets emitting higher concentrations of CO. Proper and continuous ventilation of wood fuel storage rooms is critically important to ensure that levels of CO are kept to a minimum.

10.3.5 Populations at Increased Risk

The health literature identifies several populations that are more susceptible to the impacts of exposure to air pollution. Susceptibility refers to populations that are affected by lower levels of pollutant concentrations than are the general population.

Chronic illness, such as preexisting respiratory disease, and acute conditions, such as influenza, put people at greater risk of adverse outcomes associated with fine particulates exposure. Also, some populations are more susceptible to health effects because of their age or condition. For example, pregnant women, infants, children, the elderly, and individuals already burdened by significant environmental, social and economic stressors are more likely to be adversely affected by exposure to air pollution (Pope and Dockery 2006; Brook et al. 2010; Johnson and Graham 2005; Landrigan 1998; Stein et al. 2008). While not all of these populations have been studied with respect to woodsmoke exposure, there is evidence that children as well as people with existing respiratory illness are particularly susceptible to the adverse effects of woodsmoke (Naeher et al. 2007). This is of particular concern given the lack of data on emissions from wood biomass units operating in schools and other facilities near susceptible populations—and associated health impacts.

In addition to these susceptibilities, populations in some geographic locations are more vulnerable than others to emissions because of existing pollution loads. In contrast to susceptibility, vulnerability refers to populations of individuals at risk of various health outcomes because they experience more frequent or higher levels of exposure. Where air quality is already poor, new sources of emissions are likely to have a more significant impact on health than in areas with good air quality. In addition, as noted earlier, communities located within geographic zones such as valleys, mountainous terrain, or river basins, or where temperature inversions are common, may experience higher levels of exposure because of meteorological and topographical features which allow pollutants to concentrate. The resulting public health impact will be greater among communities in these regions, especially among those with greater population densities.

10.3.6 Additional Considerations: Potential Health Benefits

In some regions of the country, wood biomass energy projects which remove flammable fuel from forest floors have the potential to reduce forest fires and associated health risks. As reviewed earlier, studies have demonstrated health problems, including respiratory outcomes as well as some evidence supporting cardiovascular outcomes, resulting from exposure to air pollution from large wildfires (Naeher et al. 2007). In 2012, a Health Impact Assessment (HIA) was conducted for a two-megawatt advanced combustion (2-stage combustion) electricity-generating wood biomass energy plant in Placer County, CA., plans for which called for fuel (wood chips, bar, and pine needles) to be supplied by local forest fire prevention programs (Sequoia 2012). The HIA reviewed several potential health consequences associated with the project, among them impacts resulting from wildfires, including air pollution. It concluded that the design of the facility would likely benefit public health through improvements in air quality from reductions in wildfires. Thus, it is conceivable that the installation of a wood biomass energy facility could result in a net reduction in air pollution, with associated public health benefits, though likely only in limited circumstances where wildfires are prevalent.

10.4 Public Health Protection in Biomass Energy Decision-Making

10.4.1 Policies for Protecting Public Health from Biomass Emissions in the Northeast: Limitations and Opportunities

10.4.1.1 Air Quality Regulation and Wood Biomass

Federal and state air quality laws, regulations, and management generally aim to ensure healthy air quality in regional airsheds, focusing on a short list of pollutants emitted from stationary and mobile sources. Under the federal Clean Air Act, EPA has established National Ambient Air Quality Standards (NAAQS) for six priority "criteria pollutants": PM_{10}, $PM_{2.5}$, NOx, CO, ozone, and lead; of these, PM_{10}, $PM_{2.5}$, NOx, and CO are of most concern in wood biomass combustion. Regulators use data from networks of monitoring stations to assess regional air quality and issue permits for stationary sources as the primary tool for controlling pollutants emitted. States with geographic areas that are out of "attainment"—meaning that air quality exceeds NAAQS standards—must submit an Implementation Plan for how they will reduce concentrations of pollutants that exceed the standards. For areas that are in attainment, state Plans must include measures to "prevent significant deterioration" to current air quality.

The Clean Air Act requires periodic review of the standards and updating if needed. Regular reviews of criteria pollutants since a 1991 lawsuit have led to lowering of standards and increasing percentages of locations considered "out of attainment." In addition to setting ambient air quality standards, EPA also establishes New Source Performance Standards designed to control emissions of criteria pollutants from various technologies (Environmental Protection Agency 2012).

The vast majority of pollutants released by wood combustion (Table 10.1) are not among the six "criteria pollutants." Instead, they are on a much longer list of contaminants called "hazardous air pollutants" (HAPs) or "air toxics."[4] In contrast to the approach taken for criteria pollutants, federal emission standards for hazardous air pollutants are based on technology available to particular sectors, rather than on concentrations of pollutants considered to pose a risk to public health. EPA has recognized industrial and commercial-scale combustion units as important sources of HAPs, and recently finalized rules based on Best Available Control Technology (BACT). The rules controlling hazardous air pollutants differ depending on whether the boiler is at a facility defined as Major Source (releasing more than 10 tons per year of a single hazardous air pollutant or 25 tons per year of any combination of pollutants) versus an "Area" source. Regulations for larger boilers tend to be more

[4] P.M. 2.5 does include some HAPs. Many toxic chemicals, such as heavy metals, acids, etc. adsorb to the fine particles surface.

stringent than for smaller boilers, and for new versus existing units (Environmental Protection 2013).

States implement their pollution control programs via a permit application process, which includes estimates of emissions based on manufacturers' specifications or stack tests, depending on the size of the unit and the applicable regulations. Typically, if a facility is expected not to contribute enough additional pollutants to result in exceedences of 24 h and annual national ambient air quality standards, if it complies with or is not subject to a New Source Performance Standard, and if it demonstrates Best Available Technology, the air permit is issued.

10.4.1.2 Gaps in Biomass-Related Air Quality Regulation

Detailed description of how federal regulations apply to biomass boilers is beyond the scope of this chapter, but several comments are important in the context of policy changes. First, a fundamental limitation in current air quality regulation is the lack of information on potential health effects of many chemicals released during wood combustion. While the six criteria pollutants have been extensively studied, there are substantial gaps in a scientific understanding of many of the hazardous air pollutants. A second fundamental limitation is that health risk assessments which underlie air quality standards focus on individual pollutants, and fail to take into account cumulative or synergistic effects of the mix of pollutants to which people may be exposed. Lastly, monitoring networks that provide data on which states depend in ascertaining air quality—and cross-checking permit decisions—are limited, particularly in rural areas.

US EPA does not require that wood biomass boilers smaller than 2.9 MWt (heat input)—which includes most of the heating systems installed in schools, and other small institutional facilities—comply with specific emission limits for criteria pollutants or for HAPS (Environmental Protection 2013). Consequently, state air quality regulators have primary responsibility for emissions from small to medium institutional/commercial wood-fired combustion units. In the Northeast, regulation varies from state to state both in terms of the size of boilers that are subject to permit requirements, and the emission limits with which boilers must comply. Figure 10.2 shows the variation across the states in the threshold size above which a biomass boiler is required to apply for a permit. For example, Maine regulates only combustion units with a combined heating capacity that is larger than 2.9 MWt, so many units installed in schools escape scrutiny. In contrast, New York, New Jersey, and Rhode Island require permits for units over 0.3 MWt. Thus, some smaller institutional combustion units, particularly in schools in some Northeast states, are not reviewed by state air quality programs, so there are no regulatory requirements or incentives for installing the cleanest burning unit.

In states that do regulate small boilers, the emission limits they impose vary (Fig. 10.3), but all are less protective than in European countries such as Switzerland and Germany. Connecticut and Rhode Island have the most protective standard among Northeast states, yet these standards are vastly less protective

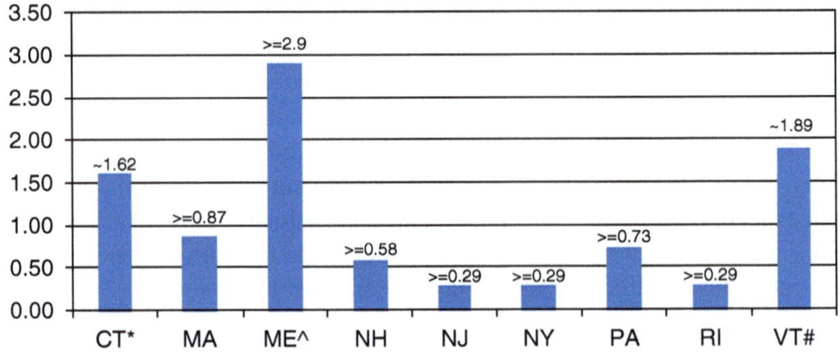

Fig. 10.2 Wood-fired combustion unit state air permit applicability threshold (MWt heat input) * - official policy, potential to emit 15 tons/yr of any pollutant. ^-combined units' heat input. # - official policy, 8.4 m^2 or more heating surface; estimate assumes 71% efficiency

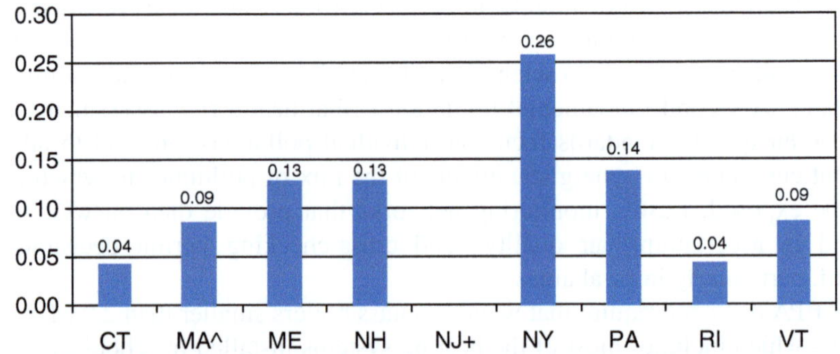

Fig. 10.3 PM emission standards for small institutional/commercial wood-fired combustion units (MWt heat input) ^ - for critical PM concern communities, + - no recent determination

than the European requirements (current European standard, 30 g/GJ) (Associates and Northeast States for Coordinated Air Use Management 2009). Thus, current PM emission limits set by the Northeast states for small-medium institutional/commercial wood-fired combustion units are neither requiring nor incentivizing innovation among boiler manufacturers in the USA to achieve what is clearly feasible based on technological advances made by European manufacturers. Moreover, inconsistent thresholds for regulation and emission limits send mixed messages to industry and fail to incentivize the cleanest burning units.

In addition to inconsistency of size thresholds and emission limits, and the failure of even the lowest emission limits to drive installation of the cleanest burning units available, the following gaps in states' air quality regulatory approaches are opportunities for enhancing public health protection from wood biomass. First, the definition of wood biomass in some Northeast states includes debris from construction and demolition, which may be contaminated with hazardous materials.[5] Second, current standardized emissions testing methods gather emissions data only when the boiler is operating at high loads and therefore do not capture the wide variability of on-the-ground operating conditions (Hoppin and Jacobs 2012). Permits may be issued on the basis of an impact analysis that used emissions data reflecting a best case scenario, rather than a typical operating period (Hoppin and Jacobs 2012). Third, lack of monitoring data for levels of most air toxics in ambient air make it difficult to identify hazardous releases. (Monitoring is limited primarily to the six criteria pollutants, but many areas do not have monitors; moreover, few states monitor for the 187 chemicals identified by EPA as "hazardous air pollutants").

Another limitation in current air quality regulation that can be addressed by state or regional policy change is the absence of routine scrutiny of risks to public health from emissions of a particular boiler or facility. Driven by ambient air quality standards in the case of criteria pollutants and BACT in the case of air toxics, federal and state air quality officials are hampered in their ability to consider the susceptibility of local populations and the vulnerability of nearby communities. Because of resource constraints, modeling is not routinely conducted on emissions data or estimates from smaller institutional boilers; only two states in the region require so-called "impact analyses," on smaller boilers, which estimate concentrations of pollutants to which people will be exposed (Hoppin and Jacobs 2012). In addition, regulations do not give authority to the states to track and take action on peak exposures which may exceed safe levels by one or more order of magnitude, nor do they empower states to specify where facilities should be sited, what fuels should be used, or boiler size.

At the program level, a major gap is training and enforcement of best practices for boiler operation including quality of fuel, boiler sizing, proper loading and avoiding operation at low loads. As noted above, emissions from a wood-fired device that are relatively clean when operating under ideal circumstances can increase substantially when fuel quality is poor—for example, when bark is burned, or fuel has high moisture content—or when the boiler unit is not well maintained (Demirbas 2008; Bignal et al. 2008; Atkins et al. 2010; Hinckley and Doshi 2010). In addition, boilers which are oversized or used too early or too late in the heating season, often operate at low loads, can result in poor combustion conditions and increased emissions of CO and $PM_{2.5}$ (Hoppin and Jacobs 2012).

10.4.1.3 Incentivizing Cleaner Technology

Federal and state policy has provided financial incentives to wood biomass under programs designed to promote renewable energy, decrease dependence on fossil fuels, and reduce carbon emissions. These range from the Northeast Regional

---

[5] Recent action by EPA has redefined facilities using construction and demolition debris as solid waste incinerators, suggesting that they will be subject to stricter emission limits. As of this writing, the rule is being "reconsidered." (Burkhard 2012)

Greenhouse Gas Initiative (REGGI), a program in the Northeast to provide monetary credits to renewable energy installations based on reduced carbon emissions, to various Fuels for Schools initiatives, which subsidize the capital costs of wood biomass boilers in schools.

Recently, health concerns about large-scale biomass combustion in Massachusetts contributed to the state's decision to restrict full renewable energy credits to facilities that achieve 60 % efficiency in combustion. (Under special circumstances, half credits will be available to facilities achieving 50 % efficiency.) (Massachusetts Executive Office of Energy and Environmental Affairs 2012) In addition, qualifying biomass energy plants in Massachusetts must emit 50 % fewer carbon dioxide emissions than combined-cycle natural gas units and also conduct a Forest Impact Assessment every 5 years to assess the negative environmental impacts of removing biomass material from forests (Simet 2012). In New York, the biomass program at the New York State Energy Research and Development Authority (NYSERDA) is providing grants for research and installation of cleanest-burning institutional and commercial-scale units. In the process, NYSERDA has developed informal criteria for cleanest-burning units, and is providing them to other grant programs interested in using the grant mechanism to distinguish truly state-of-the-art boilers that have the potential reduce emissions to levels comparable to the fuel they are replacing (Burkhard 2012).

With these exceptions, there is a general lack of incentives for the manufacture, use and distribution of cleanest-burning technologies. In addition, overt disincentives and barriers to introducing cleaner technology in the Northeast have been observed; for example, differences in US and EU testing requirements that affect a manufacturer's ability to implement a common design for biomass combustion units that can satisfy safety code requirements in both the USA and the EU. (Associates and Northeast States for Coordinated Air Use Management 2009)

Another limitation to the development of cleaner technologies is the reliance by regulators on Maximum Available Control Technology or BACT as standards for controlling hazardous air pollutants. While promoting MACT and BACT was the practical solution to the overwhelming task of regulating individual hazardous air pollutants, standards focused on end-of-stack control are inefficient at advancing new design technologies that are cleaner at their source. Best Available *Design* Technology would focus on the design of the unit itself. European combustion units would be current "best in class" from a design standpoint (Musil-Schläffer et al. 2010).

10.4.1.4 Environmental Policy Acts

The federal National Environmental Policy Act (NEPA) and parallel laws at the state level aim to ensure that the potential environmental impacts of a proposed project or "action" in which the government is involved are understood and taken into account in decision-making. Most laws define environmental impacts broadly to include human health, creating an opportunity for consideration of health effects in the environmental review process. For example, the impacts to be assessed under

Connecticut's Environmental Policy Act include the following that are relevant to biomass burning: "air quality," "the creation of a hazard to human health or safety," and "increase in traffic." (Connecticut Department of Environmental Protection 2010) If federal and state funds are involved, permits for wood-fired combustion units can trigger federal and/or state laws to assess environmental impact. However, wood-fired combustion units may not meet thresholds for comprehensive environmental review in some states.

Depending on communications and relationships among agencies in a particular state, officials responsible for implementing environmental policy acts may proactively and routinely involve public health officials in the process of environmental review from its earliest stages. At the other end of the spectrum, the agency responsible for conducting the reviews would not seek input from public health officials, placing the burden on the health agency to provide feedback at the points at which public comment is requested—either early at the scoping stage, or once the Environmental Impact Statement (or equivalent) is complete, depending on the state. As was the case in a Massachusetts biomass decision in 2011, If health officials are not involved early-on, state agencies promoting a given facility may be blindsided by health concerns raised by the community, or by the public health agency. The growing interest in HIA as a tool for public health review is an opportunity for promoting consistency in health department reviews; HIAs have influenced decisions in at least 4 biomass energy-related projects (Wernham 2013).

10.4.1.5 Environmental Justice Policies

Environmental justice policies seek to ensure that "no segment of the population, because of its economic or racial make-up, should bear a disproportionate share of the risks and consequences of environmental pollution or be denied equal access to environmental benefits." An Executive Order issued by President Clinton in 1994 urged agencies to "make achieving environmental justice part of [their] mission by identifying and addressing, as appropriate, disproportionately high and adverse human health or environmental effects of programs, policies and activities on minority populations and low-income populations." (Presidential Executive 1994) The memo accompanying the Executive Order directs federal agencies to analyze the environmental effects, including human health, economic, and social effects of federal actions, including effects on minority communities and low income communities, when such analyses are required by NEPA. EPA has issued guidance for how to consider environmental justice issues as part of the environmental impact assessment process under NEPA. The Guidance makes clear that NEPA reviews should include assessments of potentially disproportionate impacts on low income populations and people of color (Environmental Protection Agency and Office of Federal Activities 1999).

Given the higher rates of asthma, cardiovascular and other chronic diseases among people of color and low income populations, environmental justice policies are potential vehicles for considering the health of a local population when making decisions about siting facilities and using technologies that have the potential to widen health disparities. An environmental justice lens keeps questions about disproportionate impact and the concept of externalities in the policy debate.

While most Northeast states have environmental justice policies, the scope and depth of the implementation vary—from virtually nonexistent to staffed and strategic, and from programmatic to linked with enforcement activities. Several of the states have targeted permitting and/or air quality in their environmental justice programs (Connecticut Department of Environmental Protection 2010; Bonorris et al. 2004). Connecticut seeks to improve access to the permitting process by environmental justice communities, for example. Massachusetts' Environmental Justice Policy states that environmental justice populations should receive heightened scrutiny of industrial facilities and air emissions. Permitting figures prominently in the New Jersey Department of Environmental Protection's quarterly meetings of its EJ Advisory Council. Pennsylvania's Environmental Justice Public Participation Policy requires enhanced public participation in certain permitting processes, including air permits for facilities classified as major source emitters of hazardous air pollutants, and permits regarding modifications to major source facilities. This policy resulted in early engagement by environmental justice groups in Chester, PA, which in turn contributed to a decision to retrofit a coal-fired plant to reduce emissions.

Resource limitations and changes in state political priorities have limited the impact of environmental justice policies. We were not able to identify a wood-fired boiler permitting process in the Northeast in which environmental justice communities were engaged early in the process as a result of a state environmental justice policy. More systematic linking of NEPA and state environmental policy act requirements with environmental justice policies could stimulate early outreach and engagement with environmental justice communities as part of every permit process.

Together, these policies—including both gaps and opportunities—are tools for elevating public health in wood biomass decisions. Filling gaps in air quality laws and programs that implement them; providing incentives for cleaner-burning technologies; building on concepts established in environmental policy acts and environmental justice policies: these point to opportunities to ensure the protection of public health as wood biomass proliferates.

10.4.2 Future Policy Directions

Though there are promising programs to build on, no individual Northeast state has a cohesive strategy—including but not limited to air quality regulation—for systemically addressing the health effects of burning wood biomass. While most states are promoting renewable energy with goals of reducing carbon emissions, no state has articulated a goal of incorporating health considerations early in decision-making about energy strategies, thereby favoring technologies that benefit the economy, the climate, *and* public health. While energy agencies have consulted with government health experts on a case-by-case basis, no state has mechanisms in place for routine and systematic collaboration among relevant agencies. While health professionals, community residents and advocacy groups have been vocal in debates about the siting of particular biomass facilities, the public is not routinely engaged in energy planning, where they could consider information and contribute to discussions about how to maximize multiple societal goals.

Five realities about wood combustion units in particular discussed earlier in this report make such an initiative—undertaken at a regional level—compelling.

- Wood-fired combustion units emit relatively high levels of $PM_{2.5}$ that are of particular concern for public health.
- The dramatic variation in emission levels of wood-fired boilers, as well as lack of monitoring data, makes it difficult to ensure that other pollutants, in particular VOCs, heavy metals, and PAHs, are not emitted at levels that pose a threat to public health.
- Some Northeast states do not have policies in place to regulate smaller institutional and commercial-scale boilers, which are being installed in buildings where susceptible populations spend substantial time, especially schools and health care facilities.
- Recent advances in cleaner combustion technologies suggest that dramatic emissions reductions are technologically feasible, though it is unclear if emissions levels from the cleanest burning technologies will be lower than those from existing fuels. Current policy does little to incentivize adoption of advanced technologies.
- The failure to ensure that biomass energy does not increase risks to public health—both directly (e.g., upstream risks to workers and downstream impacts on people exposed to woodsmoke) and indirectly (if excessive demand for wood fuel leads to unsustainable forestry practices and carbon emissions), threatens the development of sustainable biomass energy strategies as one alternative to fossil fuels.

10.4.3 Framework for Policy Change to Address the Health Impacts of Wood Biomass

A Symposium convened by the Lowell Center for Sustainable Production in 2011 brought together representatives of advocacy organizations, health scientists, energy, environment, health and education officials, and the biomass industry to consider the most recent science on health effects and the current capacity in state government to protect public health from wood biomass. In six break-out discussions, the participants in the Symposium—70 people from all nine Northeast states—embraced a common goal of healthy renewable energy and recommended action steps that they believed held promise for elevating health in biomass decision-making and ensuring the protection of public health. The action steps, the five broad recommendations that emerge from them, and "guiding principles" that the Lowell

Center discerned from discussions at the meeting—are the foundation of a policy framework that reflects common ground among stakeholders in the Northeast. Following the Symposium, a survey of participants identified nine priority action steps that are particularly ripe for implementation (Hoppin and Jacobs 2012).

10.4.3.1 Guiding Principles

- *It is important to fill relevant data gaps, but there is sufficient scientific information to proceed with common-sense actions to reduce exposures to woodsmoke.* Symposium participants recommended both filling policy-relevant research gaps and taking action before the gaps are filled. This principle acknowledges that systems of production and consumption are not only root causes of pollution but also part of the solution.
- *Though there may be disagreement about the pace of change that is needed or feasible, the principle of continual improvement is important*: all policies should drive continual improvement in technologies to achieve greater combustion efficiencies and lower emissions of pollutants.
- *Regional coordination is needed.* Innovation can be undermined or encouraged by state policies and programs. Coordination among agencies within states and across the Northeast region—by organizations such as the Northeast States for Coordinated Air Use Management (NESCAUM), the Asthma Regional Council of New England, the American Lung Association, and the regional offices of EPA, DHHS, and the US Forest Service—is needed to maximize the effectiveness and efficiency of regulations, programs and other tools that protect public health and advance cleanest-burning technologies.
- *It is important to consider the health and environmental impacts of wood biomass technologies across the life cycle—upstream as well as downstream.* An examination of the full life-cycle supports a more complete consideration of impacts and trade-offs. That said, while a broader understanding of system-level impacts has the potential to point to solutions that avoid unintended health and environmental consequences, it is important to define a scope which is manageable given available analytical tools, which capitalizes on opportunities for promoting better—if not perfect—practices, and which encourages continual improvement.

10.4.3.2 Recommendations

1. *Constructively engage the public in wood biomass decisions*, providing opportunities for full participation in project and policy planning, including providing and considering relevant data, weighing trade-offs, and proposing solutions.
2. *Prioritize public health in wood biomass decision-making across the Northeast.* Consider public health implications in the energy planning process, maximizing health protection while also respecting other important societal goals, such as

energy efficiency, carbon neutrality, sustainable forestry practices, reduced dependence on fossil fuels, and economic revitalization.

3. *Promote a better understanding and consideration of the impacts of wood biomass on susceptible and vulnerable populations, as well as measures to prevent or reduce hazardous exposures to reduce disproportionate health effects.* Pay particular attention to the risks from localized peak exposures and the installation of wood boilers in schools.
4. *Incentivize and reward only high-efficiency, clean, wood-fired combustion,* with consistent standards across the Northeast, focusing not only on technologies but also on outcomes.
5. *Fill gaps in existing air quality regulation and air quality monitoring capacity,* including lack of regulatory scrutiny of smaller ICI units in some states. Consider other regulatory measures to protect public health and discourage all but the cleanest-burning wood biomass units.

10.4.3.3 Priority Action Steps

1. Formally *integrate health into energy planning processes by advancing health impact assessment* for proposed energy projects as well as for state-wide energy planning, engaging stakeholders early in the design and execution of the HIA.
2. *Develop standardized Health Impact Assessment methods appropriate for the broad range of energy projects.* Standardization of health impact assessment methods is needed for projects ranging from a proposed boiler in an elementary school to a 30 MW electricity-generating plant.
3. *Establish regional specifications for appliances,* including efficiency and emission standards, drawing on real performance data.
4. *Design and conduct an efficient study of the health* effects (or biological markers) to address whether children are being adversely affected by woodsmoke in their schools, including school absenteeism.
5. *Develop a best practices guide* for optimizing biomass heating combustion efficiency & performance. The guide would review state-of-the-art high-efficiency biomass heating system design and best-performing emission control technologies as well as practices for ensuring high combustion efficiencies and lower emission levels.
6. *Establish regional specification standards for wood biomass fuel* (e.g., ash, moisture content etc.).
7. *Provide incentives or financing programs to offset the up-front costs of new cleanest-burning wood biomass heating projects.*
8. *Improve understanding of the emission rates and ambient air impacts of criteria pollutants and air toxics* associated with industrial, commercial, institutional and electricity-generating wood biomass combustion, given variability in operating and load characteristics, fuel types and meteorological and topographical conditions. Because ambient air impacts are affected by source characteristics (e.g., stack

height, exit temperature and velocity) as well as meteorological and topographical characteristics, air dispersion analyses are needed for those smaller combustion units that often fall below regulatory thresholds.

9. *Establish a regional working group to integrate public health into the energy decision-making* process, building on existing effective regional organizations.

10.5 Conclusion

The scientific information summarized in this chapter makes clear that exposure to pollutants found in woodsmoke, as well as to woodsmoke itself, poses risks to health. Susceptible individuals, including children and the elderly, people with existing respiratory disease, as well as people in vulnerable communities—where multiple sources of pollution or topographical, meteorological, or other conditions tend to increase exposure—are at greater risk than the general population. This chapter has provided a framework for integrating health into decision-making about wood biomass as a renewable energy strategy in the Northeast. Applying the guiding principles, recommendations, and action steps that comprise this framework may result in decisions not to promote wood biomass in certain settings. In others, considering health alongside other important societal goals may lead to decisions to install wood biomass boilers; in these cases, bold measures are needed to ensure that only the cleanest burning units are installed, with ongoing incentives for continual improvement of technologies and practices, and disincentives for units that emit higher levels of pollutants than alternative fuels that would otherwise be used. Many organizations and sectors have a role to play in ensuring the protection of public health as the combustion of wood biomass proliferates. Representatives of federal and state energy, health, environment, education, and forestry agencies in the Northeast, the biomass industry, and health advocacy organizations have demonstrated their capacity to work collaboratively towards the common goal of healthy renewable energy.

References

Asmussen P, Duff WL, Heidtman ER et al (1986) Acute respiratory illness following occupational exposure to wood chips—Ohio. MMWR Morb Mortal Wkly Rep 35(30):483–484, 489–490

Agency for Toxics Substances Disease Registry (2010) Toxics substances portal: toxicological profiles. http://www.atsdr.cdc.gov/toxprofiles/index.asp. Accessed 15 Sept 2010

Allen R, Larson T, Sheppard L et al (2003) Use of real-time light scattering data to estimate the contribution of infiltrated and indoor-generated particles to indoor air. Environ Sci Technol 37(16):3484–3492

Allen RW, Mar T, Koenig J et al (2008) Changes in lung function and airway inflammation among asthmatic children residing in a woodsmoke-impacted urban area. Inhal Toxicol 20(4):423–433

Allen RW, Carlsten C, Karlen B et al (2011) An air filter intervention study of endothelial function among healthy adults in a woodsmoke-impacted community. Am J Respir Crit Care Med 183(9):1222–1230

Association of Occupational and Environmental Clinics (AOEC) (2010) Exposure code lookup. http://www.aoecdata.org/ExpCodeLookup.aspx. Accessed 15 Oct 2010

Atkins A, Bignal KL, Zhou JL et al (2010) Profiles of polycyclic aromatic hydrocarbons and polychlorinated biphenyls from the combustion of biomass pellets. Chemosphere 78:1385–1392

Barregard L, Sallsten G, Gustafson P et al (2006) Experimental exposure to wood-smoke particles in healthy humans: effects on markers of inflammation, coagulation, and lipid peroxidation. Inhal Toxicol 18(11):845–853

Barregard L, Sallsten G, Andersson L et al (2008) Experimental exposure to wood smoke: effects on airway inflammation and oxidative stress. Occup Environ Med 65(5):319–324

Beauchemin PA, Tampier M (2008) Emissions from wood-fired combustion equipment. Envirochem Services, Inc., Prepared for British Columbia, Ministry of Environment, Environmental Management Branch, Accessed 30 June 2008

Bellmann et al (2008) In: Joikiniemi J, Hytönen K et al. Biomass combustion in residential heating: particulate measurements, sampling, and physicochemical and toxicological characterisation. ISSN 0786-4728 Final report of the project "Biomass-PM" funded by ERA-NET Bioenergy Programme 2007–2008. University of Kuopio

Bignal KL, Langridge S, Zhou JL (2008) Release of polycyclic aromatic hydrocarbons, carbon monoxide and particulate matter from biomass combustion in a wood-fired boiler under varying boiler conditions. Atmos Environ 42:8863–8871

Boman BC, Forsberg AB, Jarvholm BG (2003) Adverse health effects from ambient air pollution in relation to residential wood combustion in modern society. Scand J Work Environ Health 29(4):251–260

Bonorris S, Isaacs J, Brown K (2004) Environmental justice for all: a 50 state survey of legislation, policy and initiatives. American Bar Association and Hastings College of the Law. Jan 2004. http://www.abanet.org/irr/committees/environmental/statestudy.pdf. Accessed 15 Nov 2010

Booth MS (2008) Letter to Marc Simpson APC, Bureau of Waste Prevention, Western Regional Office, Massachsuetts Department of Environmental Protection. 8 Jul 2008, (personal communication)

Booth MS (2012a) Biomass energy in Pennsylvania: implications for air quality, carbon emissions, and forests. Partnership for Public Integrity (prepared for: The Heinz Endowments). Dec 2012

Booth MS (2012b) Biomass energy in Pennsylvania: implications for air quality, carbon emissions and forests partnership for public integrity (prepared for: The Heinz Endowments). Dec 2012. p 61, Table 9

Brook RD, Rajagopalan S, Pope AC III et al (2010) Particulate matter air pollution and cardiovascular disease: an update to the scientific statement from the American Heart Association. Circulation 121:2331–2378

Brown DR, Callahan BG, Boissevain AL (2007) An assessment of risk from particulate released from outdoor wood boilers. Hum Ecol Risk Assess 13:191–208

Brauer M (2012) Biomass emissions, exposure and health effects. In: Hoppin P, Jacobs M (eds) Wood biomass for heat & power addressing public health impacts. Symposium Summary. Lowell Center for Sustainable Production, University of Massachusetts, Lowell, pp 18–22

Burkhard E (2012) New York Department of Energy, Research and Development Authority. 4 Sept 2012, (personal communication)

Chandrasekaran SR, Hopke PK, Rector L et al (2012) Chemical composition of wood chips and wood pellets. Energ Fuel 26(8):4932–4937

Chandrasekaran SR, Laing JR, Holsen TM, et al (2011) Emission characterization and efficiency measurements of high-efficiency wood boilers. Energy Fuels 25(11):5015–5021

Connecticut Department of Environmental Protection (2010) Connecticut Environmental Policy Act an Environmental Permitting Fact Sheet. http://www.ct.gov/dep/cwp/view.asp?A=2709&Q=324144. Accessed 15 Nov 2010

Crouse DL, Peters PA, van Donkelaar A et al. (2012) Risk of nonaccidental and cardiovascular mortality in relation to long-term exposure to low concentrations of fine particulate matter: a Canadian national-level cohort study. Environ Health Perspect 120(5):708–714

Daigler GE, Markello SJ, Cummings KM (1991) The effect of indoor air pollutants on otitis media and asthma in children. Laryngoscope 101(3):293–296

Danielsen PH, Loft S, Jacobsen NR et al (2010) Oxidative stress, inflammation and DNA damage in rats after intratracheal instillation or oral exposure to ambient air and wood smoke particulate matter. Toxicol Sci 118(2):574–585

Danielsen PH, Møller P, Jensen KA et al (2011) Oxidative stress, DNA damage, and inflammation induced by ambient air and woodsmoke particulate matter in human A549 and THP-1 cell lines. Chem Res Toxicol 24:168–184

Delfino RJ, Brummel S, Wu J et al (2009) The relationship of respiratory and cardiovascular hospital admissions to the southern California wildfires of 2003. Occup Environ Med 66(3):189–197

Demers P, Teschke K (1998) Chapter 71: lumber. In: Stellman JM (ed) Encyclopeadia of occupational health and safety, vol III, 4th edn. International Labour Office, Geneva

Demirbas A (2008) Hazardous emissions from combustion of biomass. Energ Sourc Part A 30:170–178

Dennekamp M, Erbas B, Sim M et al (2011) Air pollution from bushfires and out of hospital cardiac arrests in Melbourne, Australia. Epidemiology 22(1):S53

Environmental Protection Agency (1995) AP 42, 5th edn, vol I, chapter 1.6: wood residue combustion in boilers. http://www.epa.gov/ttnchie1/ap42/ch01/. Accessed 29 Jul 2013

Environmental Protection Agency (1995) AP 42, 5th edn, vol I, chapter 1.6: wood residue combustion in boilers (Table 1.6-1). http://www.epa.gov/ttnchie1/ap42/ch01/. Accessed 29 Jul 2013

Environmental Protection Agency (2013) National emissions standards for hazardous air pollutants for area sources: industrial, commercial and institutional boilers; final rule, vol. 78, Federal Register, 1 Feb 2013, pp 7488–7522

U.S. Environmental Protection Agency (2012) Laws and Regulations (Clean Air Act, 42 U.S.C. 7401 et seq. (1970)). http://www.epa.gov/lawsregs/laws/caa.html. Accessed 12 Nov 2012

U.S. Environmental Protection Agency, Office of Federal Activities (1999) Final Guidance for Consideration of Environmental Justice in Clean Air Act 309 Reviews. Jul 1999. Final Guidance for Consideration of Environmental Justice in Clean Air Act 309 Reviews. Accessed 15 Nov 2010

Gauthier S, Grass H, Lory M et al. (2012) Lethal carbon monoxide poisoning in wood pellet storerooms—two cases and a review of the literature. Ann Occup Hyg 56(7):755–763

Health Effects Institute Air Toxics Review Panel (2007) Mobile-source air toxics: a critical review of the literature on exposures and health effects. Summary of studies of diesel exhaust. Health Effects Institute

Henderson SB, Brauer M, Kennedy S et al (2009) Three measures of forest fire smoke exposure and their association with respiratory and cardiovascular physician visits and hospital admissions. Epidemiology 20(6):S82

Hinckley J, Doshi K (2010) Emission controls for small wood-fired boilers. Resource Systems Group Inc, White River Junction, VT, USA

Holton WC (2001) Special treatment: disposing of CCA-treated wood. Environ Health Perspect 109(6):A274–A276

Hopke PK, Holsen TM (2010) Evaluation of the energy and emissions performance of commercial scale advanced wood combustion systems. Clarkson University. Prepared for the New York State Energy Research and Development Program. Aug 2010

Hoppin P, Jacobs M (2012) Wood biomass for heat and power: addressing public health impacts. Symposium summary. University of Massachusetts Lowell, Lowell Center for Sustainable Production. www.sustainableproduction.org. Accessed 15 Oct 2012

Humphrey DN (2005) A Study for Boralex Energy, Inc.: fate of dioxin and arsenic from the combustion of construction and demolition debris and treated wood. University of Maine, Orono, ME, USA

Inspection: 313204299—Creative Biomass Inc. (2011) Occupational Safety and Health Administration. http://www.osha.gov/pls/imis/establishment.inspection_detail?id=313204299. Accessed 15 Jan 2011

International Agency for Research on Cancer (2010) IARC monographs on the evaluation of carcinogenic risks to humans volume 95: household use of solid fuels and high-temperature frying

International Agency for Research on Cancer (2013) Agents classified by the monographs, vol. 1–108. http://monographs.iarc.fr/ENG/Classification/. Accessed 29 Jul 2013

Johansson LS, Leckner B, Gustavsson L et al (2004) Emission characteristics of modern and old-type residential boilers fired with wood logs and wood pellets. Atmos Environ 38:4183–4195

Johnson PR, Graham JJ (2005) Fine particulate matter national ambient air quality standards: public health impact on populations in the northeastern United States. Environ Health Perspect 113(9):1140–1147

Karr CJ, Demers PA, Koehoorn MW et al (2009) Influence of ambient air pollutant sources on clinical encounters for infant bronchiolitis. Am J Respir Crit Care Med 180(10):995–1001

Klippel N, Nussbaumer T (2007) Health relevance of particles from wood combustion in comparison to diesel soot. 15th European Biomass Conference and Exhibition, 7–11 May 2007, Berlin

Kocbach Bølling A (2012) The toxicity of woodsmoke particles generated under different combustion conditions. In: Hoppin P, Jacobs M (eds) Wood biomass for heat & power addressing public health impacts. Symposium Summary. Lowell Center for Sustainable Production, University of Massachusetts, Lowell, pp 22–28

Krewski D, Burnett RT, Jerrett M et al (2009) Extended follow-up and spatial analysis of the American Cancer Society study linking particulate matter air pollution and mortality. Health Effects Institute, Boston, MA

Landrigan PJ (1998) Environmental hazards for children in USA. Int J Occup Med Environ Health 11(2):189–194

Larson T, Gould T, Simpson C et al (2004) Source apportionment of indoor, outdoor, and personal PM2.5 in Seattle, Washington, using positive matrix factorization. J Air Waste Manag Assoc 54(9):1175–1187

Levelton Consultants Ltd. (2010) Emissions and technology review of selected biomass conversion systems. Prepared for Nexterra Systems Corporation. http://www.nexterra.ca/PDF/101025_Levelton_Emissions_Report_Executive_Summary_Oct_2010.pdf. Accessed 1 June 2011

Levy JI, Bennett DH, Melly SJ et al (2003) Influence of traffic patterns on particulate matter and polycyclic aromatic hydrocarbon concentrations in Roxbury, Massachusetts. J Expo Anal Environ Epidemiol 13(5):364–371

MacIntyre EA, Karr CJ, Koehoorn M et al (2011) Residential air pollution and otitis media during the first two years of life. Epidemiology 22(1):81–89

Maine Department of Environmental Protection (2007) Report on the substitution of wood from construction & demolition debris for conventional fuels in biomass boilers

Massachusetts Executive Office of Energy and Environmental Affairs (2012) Massachusetts Department of Energy Resources: summary of final revisions to the RPS biomass regulation and guidelines

McDonald R (2009) Evaluation of gas, oil and wood pellet fueled residential heating system emissions characteristics. Brookhaven National Laboratories Energy Sciences and Technology Department/Energy Resources Division (BNL-91286-2009-IR)

Miller P, Allen G, Rector L et al (2010) Spatial modeling and monitoring of residential woodsmoke across a non-urban upstate New York region. Northeast States for Coordinated Air Use Management. Prepared for the New York State Energy Research and Development Authority

Musil-Schläffer B, McCarry A, Schmidl C (2010) European wood heating technology survey. New York State Energy Research and Development Authority. http://www.nyserda.org/programs/Research_Development/10-01_european_wood_heating_technology_survey.pdf. Accessed 15 Nov 2010

Naeher LP, Brauer M, Lipsett M et al (2007) Woodsmoke health effects: a review. Inhal Toxicol 19(1):67–106

NARSTO (2010) Improving estimates of air pollutant emissions in the northeast and mid-Atlantic states: a workshop report and summary. Mar 2010

Noonan CW, Ward TJ, Navidi W et al (2012) A rural community intervention targeting biomass combustion sources: effects on air quality and reporting of children's respiratory outcomes. Occup Environ Med 69(5):354–360

Northeast States for Coordinated Air Use Management (2006) Emissions from burning wood fuels derived from construction and demolition debris. May 2006

Oanh NTK, Reutergardh LB, Dung NT (1999) Emission of polycyclic aromatic hydrocarbons and particulate matter from domestic combustion of selected fuels. Environ Sci Technol 33:2703–2709

Ostro B, Roth L, Malig B et al (2009) The effects of fine particle components on respiratory hospital admissions in children. Environ Health Perspect 117(3):475–480

Pope AC III, Dockery DW (2006) Health effects of fine particulate air pollutions: lines that connect. J Air Waste Manag Assoc 56:709–742

Presidential Executive Order (1994) Federal Actions to Address Environmental Justice in Minority Populations and Low-Income Populations. 11 Feb 1994. http://www.ejnet.org/ej/execorder.html. Accessed 15 Nov 2012

Press Release: US (2011) Labor Department's OSHA cites Franklin County, Maine, pellet mill for combustible dust and other hazards after explosion. Occupational Safety and Health Administration, Region 1. http://www.osha.gov/pls/oshaweb/owadisp.show_document?p_table=NEWS_RELEASES&p_id=17235. Accessed 15 Jan 2011

Press Release: US (2011) Labor Department's OSHA proposes $135,200 in fines against Jaffrey, N.H., wood pellet manufacturer following January accident. Occupational Safety and Health Administration. http://www.osha.gov/pls/oshaweb/owadisp.show_document?p_table=NEWS_RELEASES&p_id=16364. Accessed 15 Jan 2011

Rector L (2010) Comparative emissions from small boilers. Presentation for the Heat the Northeast Conference, Manchester, NH

Richmond-Bryant J, Saganich C, Bukiewicz L et al (2009) Associations of PM2.5 and black carbon concentrations with traffic, idling, background pollution, and meteorology during school dismissals. Sci Total Environ 407(10):3357–3364

Rick Handley and Associates, and Northeast States for Coordinated Air Use Management (2009) Biomass boiler and furnace emissions and safety regulations in the northeast states: evaluation and options for regional consistency. Submitted to Massachusetts Department of Energy Resources by CONEG Policy Research Center, Inc. June 2009

Sallsten G, Gustafson P, Johansson L et al (2006) Experimental wood smoke exposure in humans. Inhal Toxicol 18(11):855–864

Schlunssen V, Madsen AM, Skov S et al (2011) Does the use of biofuels affect respiratory health among male Danish energy plant workers? Occup Environ Med 68(7):467–473

Sequoia Foundation (2012) A Health Impact Assessment of the Proposed Cabin Creek Bioass Energy Facility in Placer County, California. http://www.placer.ca.gov/Departments/CommunityDevelopment/Planning/Biomass//~/media/cdr/Planning/Biomass/Reports/HIAReport.ashx. Accessed 23 May 2013

Sippula O, Hokkinen J, Puustinen H et al (2009) Particle emissions from small wood-fired district heating units. Energ Fuel 23:2974–2982

Snook S (2012) Air emissions and permitting: ICI biomass boilers. In: Hoppin P, Jacobs M (eds) Wood biomass for heat & power addressing public health impacts. Symposium Summary. Lowell Center for Sustainable Production, University of Massachusetts, Lowell, pp 29–34

Solo-Gabriele H, Kormienko M, Gary K (1999) Alternative chemicals and improved disposal-end management practices for CCA-treated wood. State University System of Florida, Florida Center for Solid and Hazardous Waste Management

Stein J, Schettler T, Rohrer B et al (2008) Environmental threats to healthy aging. Greater Boston physicians for social responsibility and science and environmental health network

Suvendrini Lena T, Ochieng V, Carter M et al (2002) Elemental carbon and PM2.5 levels in an urban community heavily impacted by truck traffic. Environ Health Perspect 110(10):1009–1015

Svedberg U, Samuelsson J, Melin S (2008) Hazardous off-gassing of carbon monoxide and oxygen depletion during ocean transportation of wood pellets. Ann Occup Hyg 52(4):259–266
Svedberg U, Petrini C, Johanson G (2009) Oxygen depletion and formation of toxic gases following sea transportation of logs and wood chips. Ann Occup Hyg 53(8):779–787
Swiston JR, Davidson W, Attridge S et al (2008) Wood smoke exposure induces a pulmonary and systemic inflammatory response in firefighters. Eur Respir J 32(1):129–138
Truesdale RS, Cleland JG (1982) Residential stove emissions from coal and other alternate fuels combustion. Specialty conference on residential wood and coal combustion. Air Pollution Control Association, Louisville, KY, pp 115–128
Urquhart B, Boyce G (2008) Wood biomass energy: local renewable fuel for commercial, institutional and industrial facilities. Massachusetts sustainable forest bioenergy inititative, Division of Energy Resources Department of Conservation and Recreation, Executive Office of Energy and Environmental Affairs. June 2008
van Assendelft AH, Raitio M, Turkia V (1985) Fuel chip-induced hypersensitivity pneumonitis caused by penicillium species. Chest 87(3):394–396
Washington State Department of Ecology (2003) Hog fuel boiler RACT determination. Apr 2003
Wernham A (2013) HIA in the energy and natural resource development sectors. www.healthimpactproject.org/resources/body/Wernham-Energy-and-Natural-Resources.pdf. Accessed 29 May 2013
Wilson WE, Suh HH (1997) Fine particles and coarse particles: concentration relationships relevant to epidemiologic studies. J Air Waste Manag Assoc 47:1238–1249

Chapter 11
Linking Knowledge with Action: Applied Social Science Considerations to Improve Woody Bioenergy Research and Development Effectiveness

Jessica Leahy and Laura Lindenfeld

11.1 Introduction

As discussed in earlier chapters, the forest and agricultural lands in the Northeast are poised to have a significant role in meeting societal demands for energy independence and renewable energy. Woody biomass, harvested and processed sustainably, can be converted into a diverse set of energy and bioproducts. In particular, woody bioenergy holds promise as a way of augmenting or replacing petroleum- or coal-based energy sources. However, additional research is needed to encourage widespread establishment of the woody bioenergy in this region. Research can reduce barriers across the entire supply chain, from plant genetics and production, to harvesting and logistics, to processing and marketing. However, there is a long recognized gap between knowledge creation through science, and action through application of that research into practice. In order to match research "supply" provided by researchers with research "demand" by stakeholders in industry, government, and communities, new models of linking researchers and bioenergy stakeholders are currently being tested. This chapter summarizes the findings of a social considerations study conducted with stakeholders who were invited to participate in the development of a multi-state sustainable bioenergy research proposal. The study was designed to gather information from stakeholders through an online

J. Leahy, Ph.D. (✉)
School of Forest Resources, University of Maine, 241 Nutting Hall, Orono, ME 04469, USA
e-mail: jessica.leahy@maine.edu

L. Lindenfeld, Ph.D.
Department of Communication and Journalism and the Margaret Chase Smith Policy Center, University of Maine, York Village, Building 4, Orono, ME 04469, USA
e-mail: laura.lindenfeld@umit.maine.edu

M. Jacobson and D. Ciolkosz (eds.), *Wood-Based Energy in the Northern Forests*, DOI 10.1007/978-1-4614-9478-2_11, survey and three stakeholder workshops. This information was used in developing a follow-on study, currently underway, to demonstrate stakeholder involvement and improve the quality of stakeholder interactions.

Specific objectives of this study included:

- Assess stakeholder needs and knowledge that are likely to affect the viability and sustainability of an emerging woody bioproducts industry in the Northeast
- Develop a prioritized list of barriers, opportunities, research needs to woody bioenergy in the northeastern region
- Evaluate the stakeholders' awareness, willingness to engage in the research process, interrelationships, and attitudes towards bioproducts harvesting and supply processes
- Use the applied social science data collected in this study to provide much-needed guidance for research that will lead to the establishment of a bioproducts industry in the Northeast region

The results obtained through the two-part study provide important and useful data for understanding key stakeholders' views on the woody bioproducts industry, interest in collaborating, and diverse perspectives and needs.

11.2 Methods

Stakeholder survey and workshop participants were recruited by faculty involved in woody bioenergy research in the Northeast USA. The research team featured over 60 faculty who have established relationships with stakeholders interested in the bioproducts industry. Participants included business leaders, nongovernmental organization leaders, and government agency representatives. Seventy four stakeholders were invited, and 40 participated.

The participants were asked to participate in a two-part study. First, they were invited by phone and email to attend the stakeholder workshop. Upon their RSVP, they were asked to complete an online survey. The survey consisted of closed and opened ended questions to assess their opinions about barriers, opportunities, and research needs for woody bioenergy and bioproducts. Participants were also asked to share their preference for collaborating with universities and forest bioproducts researchers. The survey was seven pages long and contained eleven questions. Questions were designed to elicit participants' interest in collaborating with the project team and to determine their preferred style of collaboration.

The follow-on stakeholder workshop session drew on the data collected in the online survey and served as the basis for gathering more detailed information. Three sessions were held simultaneously at different locations: Orono, Maine (UMaine); Syracuse, New York (ESF), and State College, Pennsylvania. The total number of participants was 40 (respectively 13 in Maine, 9 in New York, and 18 in Pennsylvania). This study received Institutional Review Board approval for use of human subjects. All participants received informed consent information prior to participating in the study.

We analyzed the survey response data using the Statistical Package for Social Sciences (SPSS) software. To protect anonymity and confidentiality, identifying markers such as names, titles, and addresses were removed from the data set. We calculated descriptive statistics, such as mean, median, mode, standard deviation, and frequency distributions, for all survey items.

As with any study, there are limitations to the data collected. First, the survey sampled stakeholders invited by faculty involved in woody bioenergy to participate in the study. Certainly there are more researchers and stakeholders who might be interested and willing to participate in woody bioenergy research projects. This survey represents the views and opinions of these stakeholders and, thus, of a particular group of individuals. Future research should build on these data to assess the reliability of both the online survey and facilitated meetings. We had a high response rate, yet some of the stakeholders who participated in the online survey did not attend the meeting, while some participated in the meeting without completing the survey. These data offer valuable insight into the input we received from the participants, but it is not generalizable to a larger population. Rather, it forms the basis for future research and provides valuable input for those developing projects related to woody bioenergy and bioproducts.

11.3 Results

11.3.1 Description of Participating Stakeholders

A total of 40 participants attended the stakeholder workshops, while 44 completed the online survey. About half of the participants attended the Penn State stakeholder workshop, while the remaining attendees attended workshops at UMaine or ESF. The stakeholders who attended the meeting often held senior-level positions within their organization or agency. The most common positions were manager, director, deputy secretary, and specialist. The opportunity to shape the direction of and to be involved with woody bioenergy research attracted the attention of key stakeholders in leadership positions. There was strong stakeholder interest in woody bioenergy research and development across the entire region.

The participating stakeholders were affiliated with a diverse set of organizations that all have a role in woody bioenergy development in the Northeast. This included government agencies who will direct public policy and have a regulatory role; bioenergy and bioproduct industry organizations who are interested in feedstock supply and consumer markets; forest industry organizations who will be supplying woody biomass; trade and environmental nongovernmental organizations who represent various constituents; consulting firms who assist with research and development and business development; agricultural industry who are interested in short-rotation woody crops and equipment; and other organizations. One key to aligning research "supply" of knowledge with stakeholder "demand" for information to use in actions is to involve and listen to a balanced set of stakeholders.

Finally, the participating stakeholders possessed some knowledge or familiarity of woody biomass, ranging from extremely familiar to somewhat familiar.

The survey contained three questions designed to elicit qualitative input on challenges, opportunities, and research priorities for the forest bioproducts industry in the Northeast. The graphs represent all responses and are not weighted to reflect first, second, and third answers.

11.3.2 Challenges/Barriers

The graph in the figure below represents participants' responses to the question, *"What are the three most significant challenges facing the forest bioproducts industry in the Northeast?"* Reponses were coded and sorted into the categories represented in the graph. Economic issues constituted the single largest area perceived barrier for these stakeholders at 44.6 %. This was next followed by public perceptions (11.6 %). While economic issues stand out as the single most significant barrier, the sum of all other perceived barriers demonstrates that economic issues are part and parcel of a broader set of barriers and challenges for bioenergy in the Northeast region.

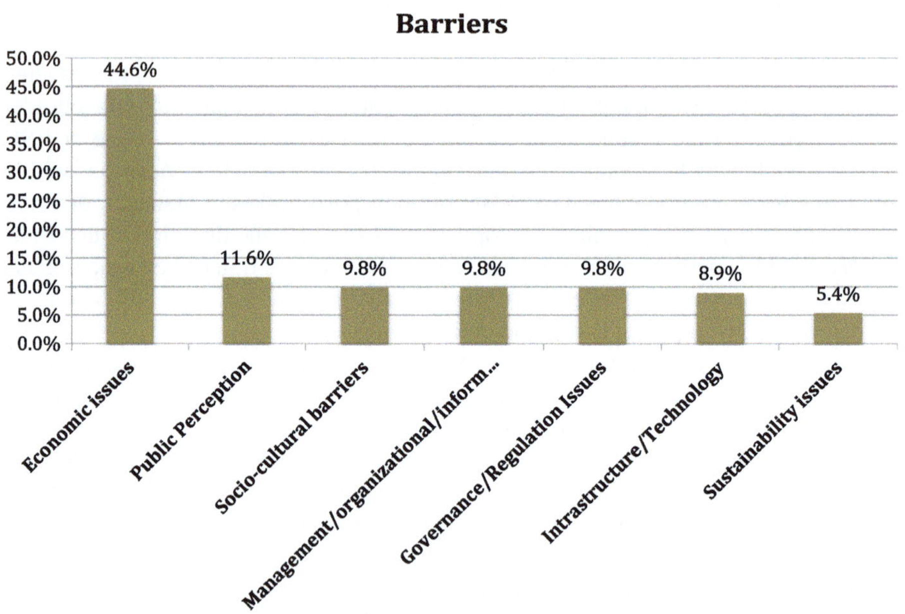

Economic issues could be divided into four subcategories:

- Supply issues such as availability of woody biomass
- Economic feasibility issues such as prices, profits, etc.
- Costs associated with logistics, especially transportation
- Market stability issues

These economic issues were incredibly helpful for the research team to know about and has been reflected in research plans moving forward.

11.3.3 Opportunities

The graph in the figure below represents participants' responses to the question, *"What are the three most important opportunities the forest bioproducts industry in the Northeast?"* Reponses were coded and sorted into the categories represented in the graph. Opportunities for building capacity based on sociocultural (especially landowner) characteristics (20.7 %) ranked as the most important opportunity for the Northeast, followed closely by the opportunity to develop renewable energy and energy independence (19.5 %). As with the discussion of key barriers to bioenergy growth and sustainability in the Northeast, the key opportunities identified by the surveyed stakeholders span across a range of issues. To be responsive to stakeholder input, it is important for research projects to address the range of opportunities presented.

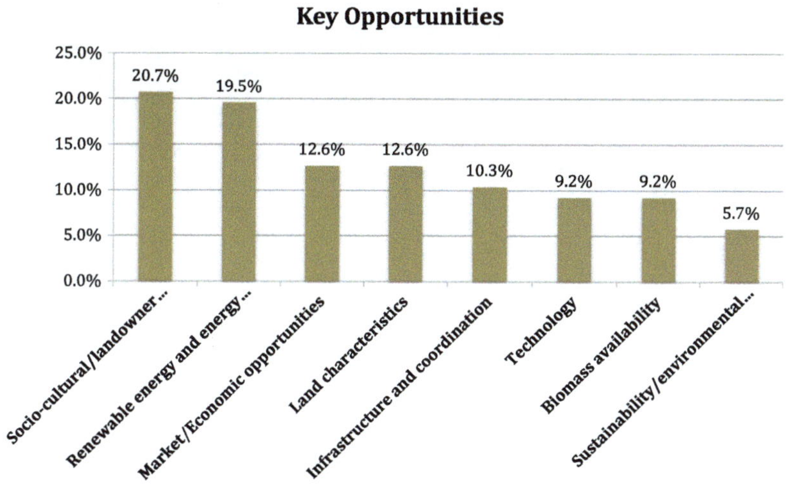

11.3.4 Research Needs

The graph in the figure below represents participants' responses to the question, "What are your three highest woody biomass research priorities for Penn State, UMaine, and SUNY-ESF to pursue in the next five years?" Reponses were coded and sorted into the categories represented in the graph.

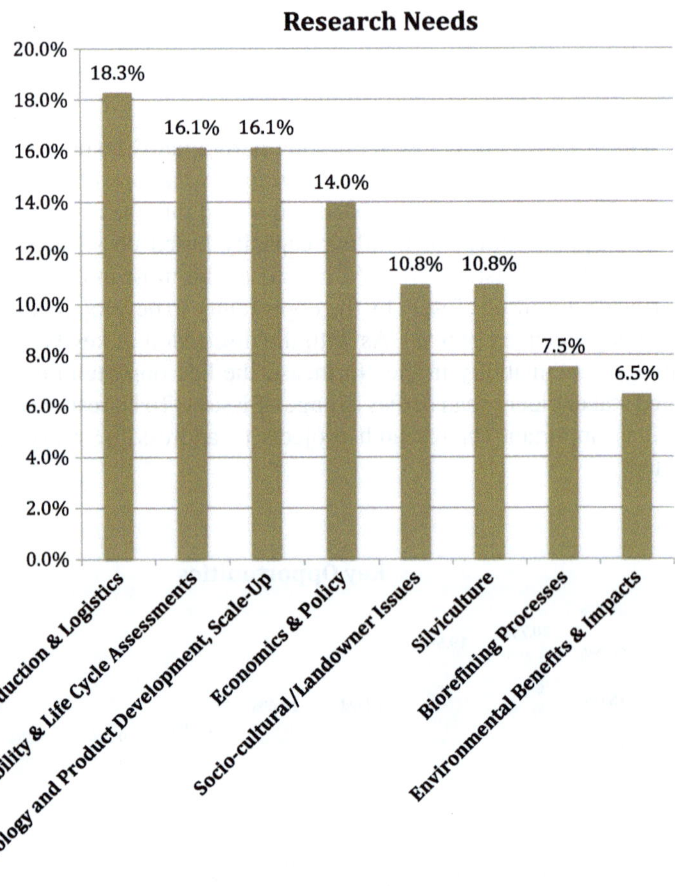

Stakeholders identified the need for diverse kinds of research. There was a clear stakeholder demand for feedstock production and logistics (18.3 %), sustainability and life cycle assessments (16.1 %), technology and product development (16.1 %). Economic and policy research (14 %) and research on sociocultural/landowner issues (10.8 %) are also notable, as is the need for silviculture research (10.8 %).

11.3.5 Stakeholder Engagement Experiences and Preferences

Discovering woody bioenergy barriers, opportunities, and research needs at the beginning of research projects is not enough in the fast changing world of bioenergy industry. It is imperative to consider how stakeholders want to be involved in the research on an ongoing basis so that there is an adaptive feedback loop between stakeholders and researchers. This helps to ensure that the research is continually aligning with stakeholder needs on-the-ground. The participating stakeholders were experienced in collaborating with faculty, staff, or students at universities on woody biomass or bioproducts research. Nearly two-thirds (64 %) of the stakeholders completing the survey and attending the meeting had worked in a collaborative relationship. Less than a third (31 %) had never been involved in collaborative research.

Among the stakeholders who had experience with collaborative research, 54 % of participants rated the experience as "very positive." Another 36 % rated their experience as "somewhat positive." Only 11 % of the stakeholders were neutral or rated their experience as "somewhat negative." Even more importantly, the stakeholders rated their past experiences in collaborative woody biomass or bioproduct research as useful to them: 61 % rated the experience as "very useful," 21 % as "useful," and 11 % as "somewhat useful." None of the stakeholders rated their collaborative research experience as any degree of not useful.

Participants rated factors that influenced their trust in university researchers working on woody biomass and bioenergy projects on 5-point Likert agreement scale (1 = Strongly disagree, 5 = Strongly agree). Researchers were most trusted because they have high technical competence, and provide reliable and useful scientific information. The lowest rated trust factor was proving timely information. Based on this, woody bioenergy research projects should focus on integration so that research, extension and education provide information in a timely manner. Supporting mechanisms for increase the speed of knowledge being put into action should be a top priority.

Stakeholders were also asked to rate their preferences for difference collaborative research models. Participants were presented with this information:

8a. Community-university partnerships can be structured in many ways. By partnerships, we mean ways university and college researchers and stakeholders can work together to address salient issues. We are interested in **your opinion** of four alternative participation strategies for stakeholders involved in community-university partnerships. In all cases, we assume implementation is handled by stakeholder.

Type of partnership	Problem Identification	Research	Proposed Solutions
A. University as **Lead** Partner	University researchers	University researchers	University researchers
B. University as **Consulting** Partner	University researchers and Stakeholder	University researchers	University researchers
C. University as **Facilitating** Partner	University researchers and Stakeholders	University researchers	University researchers and Stakeholders
D. University as **Full** Partner	University researchers and Stakeholders	University researchers and Stakeholders	University researchers and Stakeholders

Just more than half (54 %) of the participating stakeholders preferred the Facilitating model of collaborative research. In this model, researchers and stakeholders work together to define the problems and research direction, then researchers carry out the research, and then researchers and stakeholders work together to identify science-based solutions. This was closely followed by the Full model (24 %). Woody bioenergy research projects can begin using preferred engagement strategies by involving stakeholders early in the proposal development process.

8b. Participation Strategy you most prefer

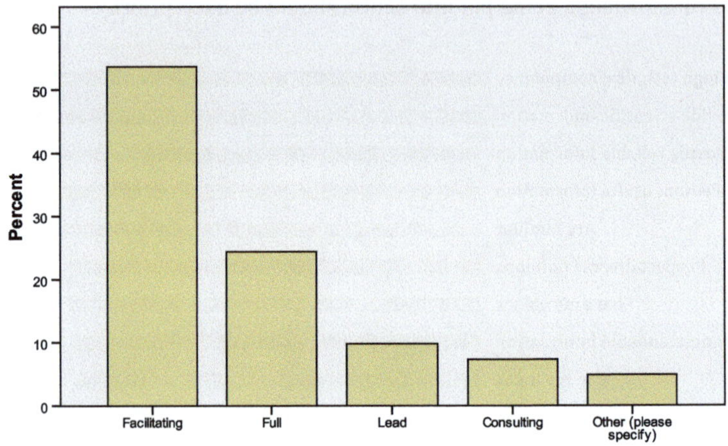

8b. Participation Strategy you most prefer

11.4 Discussion and Conclusion

Our study aimed to better understand how applied social science can successfully operate within the context of woody bioenergy research. Our goal was to identify ways to better link knowledge with action so that research "supply" is aligned with stakeholder "demand." This is especially important within the context of woody bioenergy as an emerging industry. There are many examples of traditional, linear knowledge transfer systems with limited success. Recent shifts in science include stakeholder driven, participatory research methods. Our data revealed clear barriers, opportunities, and research needs, along with stakeholder preferences for being involved in research and development activities. There were wide ranging research needs that must be addressed for industry to succeed. Based on this, projects should aim to be interdisciplinarity, capitalizing on that strength in addressing the diverse range of issues, needs, and opportunities. It is imperative that a strong stakeholder engagement system be in place to respond to iterative cycles of feedback from stakeholders and balance this with project's research capacity. Coordination and communication across research teams and stakeholders will be crucial to any project's success. Stakeholders enjoyed being invited to give input and were interested in knowing that their time and energy would result in meaningful outcomes. It is important to create a shared sense of vision and collaboration with these key stakeholders. We also learned to take the diversity of stakeholders into consideration. Communication needs are different depending on whether a stakeholder is from industry or another sector.

Research collaboratives are in a unique position to provide focus, knowledge and leadership that provides the woody bioenergy industry needed information. The participating stakeholders trust researchers and believe that woody bioenergy research produced by universities, in particular, is credible. It is important that the universities lead in this effort, but that we do so collaboratively with key stakeholders and communities. Researcher "turnaround time" is perceived as somewhat slow. Anything that can improve the time it takes researchers to bring results back to stakeholders will be viewed very positively. Collaborative efforts are necessary. As one participant stated, "*We have biomass, we have technology, we have infrastructure, we have the know how—it's about putting those together—showing people it can be done, it can be done effectively, it can be done sustainably, and move forward.*" Linking knowledge to action through demonstration sites, for instance, could be a great way to link researchers and stakeholders together at the boundary between knowledge and action. This study serves as one example of how applied social science can be helpful for identifying ways of enhancing the effectiveness and efficiency of research needed to support woody bioenergy in the Northeast.

Chapter 12
Financing for Project Development

Adam Stern

12.1 Introduction

This chapter discusses project financing, treating the reader as a project developer interested in creating a successful industrial-scale wood energy business such as a biorefinery.

Before we start "digging" into project financing, I want to share with you an interesting observation related to this book and my experience as a renewable energy investment banker. When the editors distributed the outline for the book they put the "Financing Projects" section as the last chapter, most likely, viewing it as many do as the last step of project development. While the editor's positioning may have changed on the final draft, or after reading this prologue, the fact is that I have observed the same business logic to be true in the project development cycle. As I hope you will learn from this chapter that thinking of project financing as the final step is a significant development fatal flaw and one which most likely will cause the project to be delayed or fail.

Please do not take my comments to mean that you should rush to capital sources or an investment banker to share your concept drawings. Financing sources and advisors should be resources throughout the development process and you should either have a deep understanding of the requirements of the financial markets or seek guidance along the way to put your best foot forward. Success is often a combination of what you know and who you know!

The concept of financing projects is surprisingly simple—you raise capital to build a project and then the cash flow from the project pays the investors back with an economic return. That being said, the complications come from years of projects that failed to provide the promised results for one reason or another. In other words, we are all being penalized by the mistakes of those that came before us—sorry,

A. Stern (✉)
The Gemstone Group, Inc, Wayne, PA, USA
e-mail: ajstern@thegemstonegroup.com

M. Jacobson and D. Ciolkosz (eds.), *Wood-Based Energy in the Northern Forests*,
DOI 10.1007/978-1-4614-9478-2_12, © Springer Science+Business Media New York 2013

but isn't this true in so many facets of life? However, if you keep this in mind as you read about the numerous requirements surrounding project financing, it might help you understand the perspective of the financier from which you are trying to extract capital and may better prepare you to find a balanced approach to get your project financed.

12.2 Financing Categories

While there are numerous ways to finance projects, they typically fall into two categories—corporate finance or project finance. Then within these categories there are numerous product and term options—debt/equity, short term/long term, recourse/non-recourse, secured/unsecured as well as other complicating factors such as tax incentives, multiple layers of capital, and accounting for assets. Given the focus of this book, this chapter focuses primarily on project finance debt which is what most project developers with or without a balance sheet desire.

Corporate finance is the area of finance where capital is raised on the merits of the business entity, which may or may not include the positive impact of the proposed new project. Note, however, that it will surely include any perceived negative impacts of the new project. Ultimately, investors will perform due diligence on the project to make sure it is understood, but at the end of the day, with corporate finance they are relying on the cash flows and assets of the corporation to repay the investment should the project fail.

Corporate finance is often used to finance projects when projects are too small, unproven, or the investor drives the structure in that direction. Investors prefer corporate finance structures because there is more to financially support the transaction should something go wrong or else due to the investor not having experience in project financing. A perfect example is a company trying to build its first biomass plant in the USA, albeit using European technology with several installations, and relying on renewable energy credits with no trading history for a portion of its cash flow. The investor may believe the story, but because the asset value and cash flow are unproven, they may require their investment to be structured to the corporation as opposed to the project entity directly.

Investments can be in the form of debt or equity. Equity is typically provided by the developing entity, or its investors, into the corporation with the specific authorization to fund the project entity. The majority of lenders (banks, funds, and insurance companies) finance projects using corporate debt. Larger institutions (i.e., super regional banks, money center banks, specialized funds or large foreign nationals) that have a specialized project finance group are more likely to consider other approaches.

Project finance is the area of finance where capital is raised on the merits of the project itself as if it is a stand-alone entity. In most cases this is how things are structured (for investor protection reasons). Investors rely on the cash flows and assets of the project to determine the project's merit without consideration of other corporate operations. In project financing...*the project is only as good as the weakest link in the chain!*

Because financiers rely only on the success of the project, they will seek ways to shore up the weak links to protect their investment. *For example, you would not tow your prized boat with a chain that has a cracked link regardless of how beautiful the rest of the chain looked—you would either fix it or get another chain.*

Project finance is a specialized investment area and investors are typically very experienced. To this end, they are typically more sophisticated, large financial institutions who tend to seek out larger project opportunities. If you attend a project finance conference, you will often hear the panelist saying "it costs us the same amount to diligence a large deal versus a small deal and the larger deals typically have better credit." While this attitude is common, it is not completely valid since smaller projects can be aggregated into larger projects and provides portfolio diversification. Certain residential areas in the Northeast have strong demand for woody biomass heating as natural gas and other resources are either not available or too costly. Project lenders could not possibly finance a residential project unless there was some sort of aggregated entity such as a leasing company. Regardless, since the 2008 financial crisis we live in an even more cautious financing world where lenders do not have an incentive to think out of the box and therefore seek the larger projects.

In project finance, the developing entity (the developer) invests its equity, regardless of its source, directly in the project entity. Project lenders also look to invest directly at the project level. The cash flows and assets of the project would secure the loan and the loan could be "recourse" or "non-recourse" depending on the final terms.[1] In other words, without recourse to some other entity, if the project fails, so does the investment because there is nothing or no one else back stopping that investment. There are several ways that project finance investors can insure their investments do not fail. As a final note, this chapter primarily discusses permanent financing (post construction), but part of the project may also include construction financing.

[1] Recourse loans allow the lender to go after other assets or cash flow sources (e.g., personal or corporate guarantee) in order to satisfy the loan in the case that the project defaults. Non-recourse loans only allow the lender to use the cash flow from the project and/or liquidate the project assets to satisfy the loan.

12.3 Getting Ready for Development

Many project developers are uniquely qualified to understand the technology and processes involved in the proposed project. The part where the developers often fall short is understanding the process needed to finance the project. This section of the chapter discusses (1) the project structure and its documents, (2) what is needed to attract financing, and (3) the unique operating environment of a project with project finance.

12.4 Establishing a Project Structure

The project company is typically a newly formed corporation, limited liability company (LLC) or partnership formed to hold all of the project assets including all of its rights, intellectual property, and obligations. Similar to a traditional operating company it is owned by its shareholders, unit holders or partners (depending on the legal structure) who would be considered the equity interest in the entity. In many cases, this equity interest is held by an intermediate holding company. The project company is typically a special purpose entity (SPE) which operates the project and does not have any other operations. The holding company does not typically have any other purpose other than holding the interest in the project company. The SPE structure allows for most liability to be contained by the project entity and gives equity investors an additional layer of protection. In order to be considered an investable entity, it also requires governance, internal controls, and management within the SPE to preserve its operational integrity separately from its owners. Nearly all of the operations are documented in one form or another in a series of agreements.

12.5 Agreements and Terms

This is where it starts to get more complicated. Depending on the type and complexity of the project, there are a series of documents which may be required. *Outside Project Agreements* include all contracts related to the development, construction ownership, and operation of the project with third parties. *Inside Project Agreements* include those agreements related to intercompany transactions, operations and maintenance agreements, technology services agreements, or administrative services agreements among partners or affiliates (Wilson & Rosati).

Here is a list of agreements related to many bioenergy related projects:

- Power Purchase Agreement
- Interconnection Agreement
- Engineering, Procurement, and Constructions (EPC) Agreement

- Supply Contracts (Feedstock Agreements)
- Sales Contracts (Off take Agreements)
- Environmental Attribute Agreements—i.e., Renewable Energy Credit (REC) Agreements
- Operations and Maintenance Agreement
- Administrative Services Agreement
- Technology License Agreement

These Project Agreements become the legal foundation of the project, specifying what may be done, who will be involved, and how much it will cost. Project financing will have its own documents including loan agreements, forbearance agreements, collateral agreements, security agreements, guarantees, easements, consents and others. The overriding principle that needs to be understood is that the project finance lender will be the senior lender and as such must control the deal in order to protect the investment. There are numerous project finance law firms that can guide a project developer through the appropriate options—here are a few key terms that are important to consider and understand prior to setting out to seek financing:

1. Consents—Lenders will require assignment of collateral to them to protect their investment. Any provision in the above reference documents that prevents such assignment will prevent the financing. In an effort to expedite the financing process, all project agreement documents should include automatic consents in the course of financing.
2. Contracts—the commercial terms of the contracts are extremely important. The more you can lock down your revenue and expense in a contractual form the more likely the lender will deem the project as financeable. The fact that you may do better if the "commodity" price fluctuates actually works against you when financing because all the lenders see is the potential weak link of the chain—e.g., feedstock pricing escalating which in turn lowers your margins and cash flow. In the likely event that not every revenue and expense can be guaranteed with a contract, developers should be prepared to support their financial projections using real world demonstrated situations such as long-term market pricing history or financials from similar plants in other parts of the world. It may also be necessary to shore up this link of the chain with a form of credit enhancement such as a feedstock or off take pricing hedge.
3. EPC Risk—Since the project and only the project is the underlying asset, if it is improperly built and therefore does not perform as projected, everyone is likely to have problems. Therefore the EPC contract and its terms are critical to a successful financing. The lender will look very hard at the credit worthiness and experience of the EPC contractor so you might as well do the same. The lender will want a "full wrap" with liquidated damages so it has a single point of contact to go to if there are problems. In other words, the lender prefers to have one, credit worthy, organization *wrap* together *all* of the underlying project agreements under itself and negotiate a damaged clause with that one counterparty in the event something during the construction goes wrong. These agreements should provide warranties and performance payments in the case of schedule

delays, cost overruns, inefficiency, equipment failures and performance defects. How these warranties work and at what level will depend on the specifics of the project and the comfort of the lender.
4. Change of Law/Force Majeure—Investors do not like the unknown or the unexpected. However, many contracts that involve utilities or large corporations as the counterparty often seek provisions that terminate the contract if a law or other event occurs that is contrary to their economic position. Obviously, if the contract is void at some future point in time, but the financing is not fully repaid, the borrower and thus its project finance lender are in big trouble. This is typically a hotly negotiated provision and one which project finance law firms are extremely qualified to address.

12.6 Due Diligence Preparation

As important as understanding the documents and terms, one needs to understand what it will take to get investors interested in your project. A financing package is a marketing package with projections of financial results. Most importantly, everything in a financing package should be fully supportable either by fact or by justification of your assumption. Due diligence is the process where an investor researches the proposed project to see if it is worthy of investment. Below are some key areas to cover in the financing package, as they will be the basis for due diligence preparation.

1. Project Description/Technology—Any financing from a sophisticated investor will require an internal analysis of your project—"a project write-up." A critical part of that analysis will be their understanding of the project. Therefore, if you want to put your best foot forward internally within their organization, you might as well do their work for them by preparing a clear and concise description of the project. The description should include details of the technology and specifically support that the technology works. The more independent engineering certifications (with diagrams) and documentation of actual operational experience (large scale plant—not a bench or pilot plant) the more the project will stand out. Plant yields, operational statistics, costs, etc used in the description should flow to the financial projections and be supportable.
2. Predictable Cash flow—At the end of the day, the investor is looking to get their money back and get a profit on their investment. Your business model should realistically show investors how this will happen. Do not over complicate the analysis. Review your business plan like you are the investor (i.e., would you write a check for $20 million to your project?) To do this effectively, look at the key drivers to your business and make sure those assumptions are predictable. For example, you should have ample resources for your feedstock and long-term contracts to ensure predicable pricing. The same goes for your sales or the offtake. Anything less provides anxiety for the investor caused by risk of the unknown. Inputs such as commodities with a fluctuating price are a risky proposition for an investor. Look at very long-term pricing trends to justify

your data inputs and then try to build in hedging products to strengthen your business model.

3. Budget/Forecasts—Set up a model that shows investors the key financial drivers of the project and illustrates how the project will generate enough cash flow. Be prepared for the investor to run significant sensitivity analysis. Knowing this, you should run those sensitivities ahead of sharing the data with investors so you know how your model will react. Submit a model to investors that is realistic or conservative, but still works under a stress test. Nobody likes the "hockey stick" version, but a rock bottom conservative version is dangerous too. It is called "hockey stick" because if you graph out the financial results it is often in the shape of a hockey stick—flat in the beginning with a sudden and often unexplained increase at the end of the project period.
4. Environmental site evaluation—Project development in general today is complicated and even more so in the energy sector. Any sophisticated investor will have an environmental checklist in their due diligence package. Therefore, as a developer, you will want to make sure that you have done the proper preparation work. Not only checking the site itself for past environmental issues, but also proving to investors that your project will not create additional environmental hazards. Clearly, you want to make sure that direct process hazards are addressed, such as proper storage of chemicals used in your project. However, you also need to be prepared to address potential environmental hazards that are "further afield." For example, wind farms need to have an avian study to insure safety of native birds and biomass projects may need a variety of forestry and wildlife reviews. Investors want to see this work completed or planned so that ecological issues do not delay the project or add unexpected costs.
5. Land surveys/lease agreements/operating agreements—make sure that these documents are completed and organized. The easier you make it for the investor, the easier it will be for you. If you cannot organize some paper, how can you build and operate a multi-million dollar project?
6. Insurance—Make sure you have insurance and it is ample to satisfy the investor. Your attorney and insurance agent will help you determine those limits. Just as importantly, make sure that your EPC contractor, suppliers, and other related parties have their own coverage within limits required by the investor.
7. Government or third party permits—Depending on the stage of the project, these permits may be received or simply applied for. Nevertheless, investors will want to verify that you have the correct approvals to proceed. Again, do your homework to make sure you have everything covered and get this documentation organized.
8. Other financing commitments, tax credit support, or grant agreements—If you have financial or operating support from others, make sure it is in writing and include this in the package. This may be challenging with some governmental grant programs that do not publish formal grant agreements—especially those that employ online systems, but investors will need proof. There is almost always some published literature, email receipt or letter available. Anything short of this is a red flag.

Proof of these items will either be required during the due diligence process or required as a condition of closing the financial transaction.

12.7 Loan Types

After a project is approved for financing, the next step is to provide the capital to construct the project, commonly referred to as construction finance. Lenders in most cases will require some project equity but really are relying on your ability to pay them upon commercial operation. These lenders must be convinced your project can be built on time and on budget, within set aside contingencies, and that the permanent capital is committed. There are a variety of construction finance options but they typically fall into something that works like a line of credit or construction term loan. The line of credit is a committed amount of capital which is drawn for certain approved reasons (i.e., to build the plant) and is due at a specific future date. Usually interest only is due while the loan is outstanding and the principal payment is due on or before a certain date. The construction term loan also provides a capital commitment and may be borrowed in periodic amounts, but the difference is that once it is borrowed, principle and interest are due on a set amortization schedule as set forth in the loan documents. This requires higher repayment amounts relative to the interest only line of credit and therefore careful consideration must be given to the construction schedule, EPC payment schedule, and permanent financing closing.

A project's permanent capital structure is more an art than a science. Depending on the specifics, project equity (including grants or other tax incentives) is typically 20–50 % of the total required capital with project debt making up the difference. A good financial model will help you determine a starting point when negotiating terms, but the debt providers will ultimately dictate the final structure. Please understand, lenders rarely give you the answer. It is a series of conversations or negotiations that allow you to settle on a structure.

Project equity will come from the owners and, especially in renewable energy projects, includes the tax equity, sponsor equity and any grant capital. The remaining project debt is commonly provided by a combination of the following types:

1. Syndicated term loans and bonds—the bulk of project debt is commonly held by a "senior" term loan or "senior" bond. These debt instruments differ primarily in the length of repayment term. They are typically used to repay construction debt and settle any outstanding construction costs. Once borrowed, the project will be required to repay principal and interest as provided in the loan documents. In smaller projects the financing may come from one lender or a small group of lenders, but larger projects typically require a syndicate of lenders, hence the term syndicated term loan.
2. Mezzanine or Subordinated loans—the senior lender may not be happy with the level of equity and the project sponsor may not have enough money or cannot raise additional money for additional equity. In these cases there are specialty lenders that offer a middle product—like the mezzanine in a building between 2 floors. This lender is senior to the equity but subordinate to the senior lender—i.e., in between the two. There are many ways to structure the repayment of the loan and terms are typically negotiated. The financial model for the project will help you and the lender determine an appropriate structure. What is important to

understand is that this capital is often priced more like equity than debt given the subordinate position. In today's terms when senior bank debt is 4–8 %, subordinated debt might be priced at 12–18 %.

3. Working capital/Letters of credit—Often the senior project lender will carve out a portion of its debt to be used as a working capital line of credit. Similar to the line of credit in the construction phase used to build the project, the lenders understand that the project entity may have short-term borrowing needs. For example, the project may need to purchase feedstock twice a year in advance of producing any product. These are special situations that are better served by something other than a term loan or bond—it does not make sense for the project to borrow this money today when it's not needed today. The project sponsor and lender often allocate these costs to a line of credit which then has special borrowing limits and rules associated. Lenders may also require reserves (cash put aside for specific reasons or in case of issues) to be built into the project cash flow "waterfall" which first go to paying down the line of credit's outstanding balances and then eliminating the need for the line.

Many renewable projects are challenging to finance because so many of the projects have unproven long-term operating results. However, at the time of writing of this chapter, there are helpful government loan and loan guarantee programs, many of which are offered by the United States Department of Agriculture. These programs are not a panacea. They often still require you to find a lender, have lots of processing involved and in the case of the guarantees, are usually not a 100 % guarantee so the lender still has to protect itself against a full loss.

12.8 Operational Restrictions

Project finance almost always requires the cessation of control at least at some important level. While many developers or sponsors do not like to give up control, it is a necessary means to an end. Keep in mind that you need the capital and investors need to make sure that their investment is protected.

In the case of equity investments, the stockholders become owners and as owners they will put in financial and operational controls to protect their investment.

When a project is financed, less obvious controls get interjected by the project finance lenders. Think of it like this…since they do not directly own the project assets and therefore would not otherwise own the cash flow, they need put contractual terms in place to insure access to the project cash flow. As such, project lenders will utilize operational and structural restrictions to insure they receive the projected cash flow in order to service their debt and maximize asset value. In addition, if something unexpected happens, these events will trigger other requirements. Failure to adhere to these restrictions will be an event of default which has contractual repercussions including but not limited to the immediate acceleration of the loan, requirements for additional cash equity investment, or notices to lenders.

(a) Reserves—project entities in most cases will be required as part of the closing to post segregated cash reserves. Reserves can be set for one or more specific items and just sit there until needed. Moreover, in many cases, if reserves are needed, there is typically a funding mechanism which requires any new "free cash flow" to first replenish reserves before it can be used elsewhere. Free cash flow may have many levels and would be a defined term in the loan agreement but for simplicity sake, one can think of it as cash flow after normal operating expenses and debt service is paid.
(b) Security/collateral—Do you remember back to buying your first home or car when borrowing money to do so? Do you remember how proud and excited you were to say you owned your first home or car? Well guess what, the lender has a first lien on that property and therefore technically owns it until you pay it off. The property is the lender's security or collateral in the event you do not pay back your loan. The same concept applies to project finance. The owner does not fully own or control the asset until the senior lenders are paid back.
(c) Lockbox/servicer—In many cases, the lender will set up a cash collection mechanism in order to insure it gets paid first with all incoming cash to satisfy its funding requirements (debt services, reserves, etc) and then the remaining cash will go to the project owner. This is typically done through a bank lockbox or through a loan servicing company.
(d) Affirmative covenants—Think of this as stuff that you are contractually obligated to do. For example, loan documents will almost always require the borrowing entity to report quarterly and annual financial statements.
(e) Financial Covenants—These are requirements to maintain certain financial ratios such as debt/cash flow or debt to income.
(f) Negative covenants—Think of this as stuff you are contractually obligated not to do. For example, loan documents will almost always prevent borrowing entities from selling or merging assets or borrowing additional money without the approval of the Lender. Again, these are ways of controlling the value of the assets and insuring that the existing loan gets paid as projected or sooner.
(g) Distribution of cash flows—Unlike corporate finance where cash flow of the company is movable across the corporation, project cash flow is carefully allocated in the project documents. The planning of where the cash can be spent as it is received is often referred to as a "waterfall." Essentially, the cash comes in and gets spent on a documented list of uses in a prescribed order of priority—see Fig. 12.1 as an example. Whatever is left at the end of the list belongs to the owner.

12.9 Implications to Project/Capital Structure

As may be gleaned from this chapter, project finance is a specialty branch of the finance sector and requires a great deal of documentation. There are even more implications associated with biomass transactions, relative to most project types. For example, the wood is a feedstock into the biomass project and it so happens that

Fig. 12.1 Project Waterfall

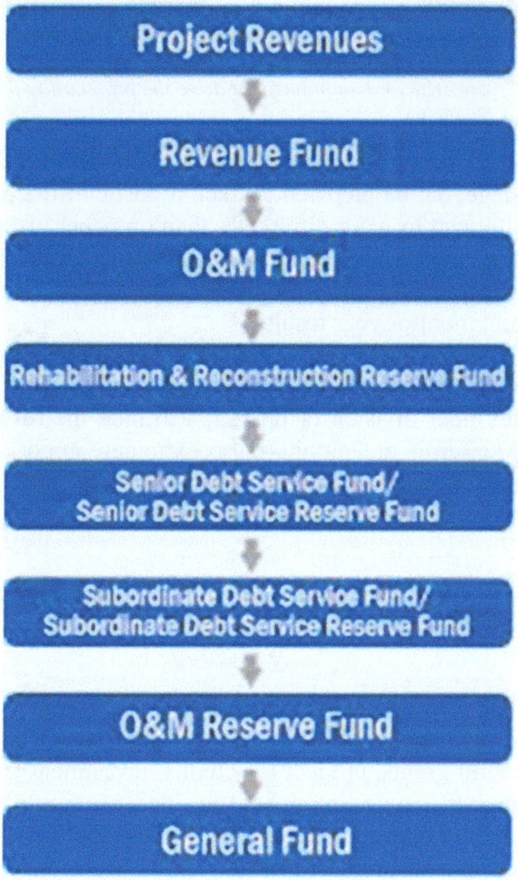

this feedstock source can also be an environmentally protected natural resource. Financiers will go to great lengths to have you document an ample supply of your feedstock (typical in most projects), but also extend this research to ensure it is not reliant on unattainable sources such as protected forests or endangered species.

12.10 Monitoring/Measuring Performance

If you think the financing hard work is over when the project is financed and built, think again. In many cases, that is when the hard work begins. When conceiving the project and developing the financial drivers of the plant, confirm that you will can track and report on those business drivers.

For example, according to statistics in a Year 2000 National Renewable Energy Labs study, a referenced small biomass plant developed to process wood waste into electricity burned nearly 400,000 pounds per hour to generate steam upwards of 900° in order to generate

30 MWe (megawatts of electricity). This is one project with one boiler technology and one feedstock source. Other projects had different statistical footprints. These important relationships between the feedstock or fuel source and the eventual offtake or energy production are critical data points that drive the profitability of the plant and one's ability to obtain financing.

These are the items on which you will run sensitivity analysis in the financing stage, but be prepared to track it post construction. Not surprisingly, things in real life tend to work differently than on paper or in demonstration pilot plants. There are "mountains" of moth balled plants from all industries that failed because of unexpectedly low performance. This is even more reason to be prepared to measure and monitor your results.

While this reporting to bankers is certainly finance related, it is reliant on operational processes to gather and communicate the data to the financial team. There is no other division or operation to hide the results, the project stands on its own. Measuring and monitoring is extremely important so that operators are able to run the plant at maximum performance. It could mean the difference for financial viability and one's ability to expand.

12.11 Tax

The US Federal and many state governments provide income tax subsidies to encourage development of renewable energy projects. These incentives include tax credit grants, product tax credits, investment tax credits, accelerated depreciation, property tax exemptions, sales and use tax exemptions and tax abatements. This chapter does not address any of these individual incentives since incentives tend to change on a regular basis.

Some incentives, like a sales tax exemption, are fairly straight forward. Others, like the federal 30 % federal investment tax credit, have complicated rules as to who is eligible and how the credit is earned. What is important to know is there are established mechanisms to take advantage of these credits and experienced professionals are required.

These incentives almost always are impactful to the economic viability of the project. Structuring to monetize these tax benefits is important but so is the administration of the benefit. Plan for the benefit, but also plan for the cost to administer. For example, I have seen several recent solar projects in one northeast state that counted on a property tax exemptions only to find out the state exemption required a local ordinance, from the individual town, which was not provided. Without the tax exemption, the economic benefit from the energy savings was nullified. In this case the project lender was fine, but the owner lost its cash profit.

12.12 Operations and Maintenance

It costs more than you think to operate and maintain a project, and costs are certain to rise over the projected life of your project. Whether you believe this statement or not, it is certainly what your investors will be thinking. Biomass plants have moving parts and in some cases chemical processes that are core to the operation. Make sure that you account for maintenance down time, cleaning, part swaps, backup part inventory, insurance limits, logistics, storms, and staff turnover/training to name just a few issues.

12.13 Conclusions

It always takes longer time, costs more money, and is more complicated than it should be, but a good project will find its money!

Reference

Wilson Sonsini Goodrich & Rosati's "Project Finance for Renewable Energy and Clean Technology Projects"

Index

M. Jacobson and D. Ciolkosz (eds.), *Wood-Based Energy in the Northern Forests*,
DOI 10.1007/978-1-4614-9478-2,

MIX
Papier aus verantwortungsvollen Quellen
Paper from responsible sources
FSC® C105338

If you have any concerns about our products,
you can contact us on
ProductSafety@springernature.com

In case Publisher is established outside the EU,
the EU authorized representative is:
Springer Nature Customer Service Center GmbH
Europaplatz 3, 69115 Heidelberg, Germany

Printed by Libri Plureos GmbH
in Hamburg, Germany